D. Elixmann · K.-H. Neumann (Eds.)

Communications Policy in Europe

Proceedings of the 4th Annual Communications Policy Research Conference, Held at Kronberg, FRG, October 25-27, 1989

With 34 Figures

Springer-Verlag
Berlin Heidelberg GmbH

Dieter Elixmann
Dr. Karl-Heinz Neumann

Wissenschaftliches Institut für
Kommunikationsdienste GmbH
Bahnhofstraße 1
D-5340 Bad Honnef 1, FRG

ISBN 978-3-540-52875-3 ISBN 978-3-540-52875-3 (eBook)
DOI 10.1007/978-3-540-52875-3

2142/7130-543210

Preface

The European Communications Policy Research Conference has taken place the fourth time since its founding in 1986. The main objective of all conferences held so far was to be a forum where social scientists, economists, engineers as well as lawyers come together to debate about current research in the field of European communications policy.

The last three conferences were organized by the Centre for Communication and Information Studies (CCIS) in London and its director Nicholas Garnham. At the 1988 Conference all participants agreed that the conference should have a future and that in particular the original ambition of the conference should be realized more fully by making it a truly European event. In the meantime, it has been proven to be possible to host the conference on a circulating yearly basis in various European countries. The Organizing Committee for this year's conference contained members from The Netherlands, France, UK and West-Germany. This year's conference was hosted and organized by WIK and held in Kronberg, West-Germany.

In many respects the past conferences had already brought about a tradition which was followed with this year's conference. There was a limited number of about 40 participants (27 coming from universities or research institutes, 3 coming from consulting firms, 8 coming from ministries, carriers or regulatory authorities, and 3 coming from supranational organisations (OECD, EC)). Attendance and presentation of a paper was by invitation only. There were only plenary sessions and no concurrent sessions. These organizational features allow both a thorough, in-depth discussion of the issues presented in the papers and a very friendly, open-minded personal atmosphere.

Relying on these concepts, this year's conference like its predecessors was an interdisciplinary event with an outstanding level of discussion between researchers and policy makers. So, the present conference volume should be of interest to academic researchers, regulators, communications industry specialists, and communications policy practitioners alike.

Even if the focus of the conference was on (West-) European perspectives on communications policy issues this did not preclude speakers and attendants alike from abroad. This year there were participants from Japan, the USA and for the first time from an East-European Country, Hungary.

This volume contains the papers given at the conference and the comments made by the discussants. After an introductory chapter the papers are organized into four basic parts: Tariff and competition issues, International and industrial aspects, Regulatory issues, Convergence issues in communications policy. There are both theoretical and empirical papers, there are international comparisons and more country specific papers, and there are papers focusing on recent developments as well as those concentrating on a more far reaching time horizon. It is our hope that this collection of papers will provide useful insights into current communications policy issues and that it contributes to this important area of interest.

Acknowledgments

The 4. Annual European Communications Policy Research Conference to which this volume refers was held October 25 - 27, 1989, at Schlosshotel Kronberg, Kronberg (Taunus) in West - Germany. We are deeply indebted to many institutions and individuals which helped us in planning, organizing and conducting the conference as well as in preparing the conference volume. Limit of space prohibits to name them all.

The conference has received financial support by the Netherlands Ministry of Education and Science (O&W, DGWB) and the Netherlands Ministry of Transport and Public Works (V&W, HDTP), the UK Economic and Social Research Council (ESRC), the French Ministry of Posts and Telecommunications, and the German Ministry of Posts and Telecommunications, which is gratefully acknowledged. Special thanks go to Jens Arnbak, Jean-Pierre Chamoux, and Nick Garnham who as members of this year's Organizing Committee helped in many ways to make the conference to an event to all participants. We would also like to express our appreciation to our staff at WIK in Bad Honnef which at the various stages helped us hosting the conference and doing the editorial work. To two individuals we owe special thanks: Walter Petry and Birgit Strüver have done an excellent job in administrating the conference and its funds; Birgit Strüver also has managed the technical details of the final processing of the conference volume. Finally, we would like to thank Springer-Verlag for incorporating the present volume into their publishing programme.

Contents

7 Multinational Banking, the Theory of the Division of Labor and the Information and Communications Technologies

Comment Chapter 7

Part IV Regulatory Models and Policies

8 Changing Regulatory Policies: The Danish Case

9 Chances and Problems of Liberalization of Telecommunications in Hungary

10 Regulatory Instruments and EEC Powers to Regulate Telecommunications Services in Europe

Joachim Scherer

Comment Part IV

Part V Convergence of Broadcasting and Telecommunications

11 Diverging Regulatory Scenarios for Integrating Broadband Communications

Emilio Lera

12 The Convergence of Broadcasting and Telecommunications in Japan

Tatsuro Hanada

Part I

Introduction

1

Introduction: Current Issues in European Communications Policy

Dieter Elixmann

Wissenschaftliches Institut
für Kommunikationsdienste
Bad Honnef

The focus of this book is on European perspectives in communications policy. Its main emphasis lies on economic issues in this field. To make more precise what we intend to present with this volume three further clarifications should be made.

Concentrating on economic topics does not mean that we believe the many different technical, legal, social or political aspects to be less important. Rather, it mirrors the objective of the conference makers to deal with the current state-of-the-art of communications policy in Europe primarily from the viewpoint of an economist. As can be seen from several contributions in this volume this does by no means preclude that other than pure economic arguments enter into the discussion.

Discussing communications policy issues relevant from a European's perspective in many respects makes it necessary to quit the national or even the European standpoint and to take into account non-European systems and current developments therein. Particularly in communications policy it becomes more and more obvious that often a global, world-wide perspective is the only reasonable one.

As indicated by the title the discussion in this volume will not exclusively refer to arguments within the scope of telecommunications policy issues but explicitly wants to make reference to the broader notion of communications policy.

To come to grips with what we think to be of special relevance and of interest concerning European communications policy perspectives on the verge of the 1990's we will give in the next section a brief overview of some recent characteristic developments in the world. An analysis of these characteristics has given rise for the selection of topics of the conference. Finally, an introduction to the papers is given.

1. Characteristics of Recent Developments in Communications Policy

Although it seems to be too early at the end of a decade to turn to a historic perspective on the passing decade we believe it to be undoubtful that there have been at least four characteristic developments during the 1980's, which shaped the decade and which will have significant spill-overs for the 1990's:

(1) A fundamental and rapid technological progress,

(2) A substantial increase in the internationalization of markets and firms' activities

(3) A reorientation towards the competitive forces of the market,

(4) A clear-cut vision of a unified Europe.

It is obvious that these factors are not completely separate and autonomous, rather there are links and repercussions simultaneously as well as through time. Notwithstanding the repercussions, each of these factors describes a different impact on the market for communications services. Thereby, both the needs for communications services and the way these services are provided have been affected.

Fundamental technical changes are obvious in particular with respect to the technology necessary to establish communication links. Indeed, both transmission and switching facilities and procedures as well as terminal equipments have undergone substantial changes. These changes relate e.g. to the digitization of the telecommunications networks, the fiber optic technology, new satellite technologies, and all kinds of mobile radio applications. Thus, technological progress allows an upgrading of already marketed services and, primarily important, the provision of completely new services.

Globalization of markets as well as perspectives of firms is apparent for the communications sector regarding e.g. the vivid activity concerning mergers and aquisitions as well as co-operation and joint ventures on the communications equipment manufacturing side. Economic activities in this field are no longer determined on a national level, rather, acting and surviving with respect to the conditions of international competition is more and more firm's strategy. The market for equipment more and more has become a world market.

Likewise, from the perspective of the provision of services increasing international co-operation between carriers is to be observed. The era of the entrenched national carrier(s) providing services in a more or less highly regulated environment according to national tastes and preferences as well as national social and political objectives is irrevocably passing. International co-operation between carriers is on the agenda sometimes leading to bundling of activities under a common roof of a newly established firm and participation of several national carriers as shareholders. Thus, geographical borders do no longer play a genuine restriction for the horizon of the economic activity of a national carrier.

Internationalization and diversification of the carrier's firm perspective is due on the one hand to the needs of consumers of communications services. Primarily those consumers for whom communications is an intermediate product, nowadays do have an increasing need for more complex and more differentiated services on an international level. On the other hand, an incentive for going international of carriers stems from the logics of a profit maximizing firm: particularly for some new services, the provision of which has been made possible by the technological inventions brought about by the 1980's (see above), there are estimates for an unprecedented economic potential of earning money allowing the carrier's financial status to become more and more independent from their former "cash cows" like plain old telephony. Thus, supporting and assuring economic viability calls for product differentiation. The internalization of these profit opportunities, however, heavily draws on expertise which can be gained in the least cost effective way by international co-operation.

The new focus on the forces of the market and the rising belief in the dominance of the market system over other more state led systems has occurred throughout the main industrialized countries of the world. This reorientation basically has its roots in the political sphere: in many Western countries election results of the 1980's have brought about a "conservative revolution". Notwithstanding the power and strength of this "revolution", which basically is an evolution, we are witnesses of real revolutions turning to the historic upheavals of the peoples of the former socialist bloc in favor of democratization and restructuring who are voting for the introduction of market economy because the complete failure of the planning economy has become apparent for them.

In Western countries this revitalization of the virtues of competition has had its impact in the field of communications policy primarily with respect to current regulation schemes. Deregulation, increased competition, efficiency, liberalization etc. have become major objectives of regulatory policy. These changes of the objectives of regulation has its effects both on a national and an international level. On the national level there are changing requirements for regulation affecting e.g. prices, quality of services, services to be offered, firms to be regulated (i.e. the contents of regulation) and the institutions and responsibilities (i.e. the organisation of regulation). On an international level there is an increasing need for harmonization of regulatory requirements (e.g. standardisation) among different countries.

The process of economic, social and political co-operation between EC member countries has reached a new dimension with the envisaged objective of "Europe 1992" aiming at a far-reaching convergence of the respective economic, social and political systems. Completion of the internal market by 1992 and a substantial transfer of political and legal power from national to EC authorities is on the agenda. Nearly all European politicians are dreaming Europe 1992 to be a real "Pareto improvement" making every member state better-off and leaving none of them worse. Substantial productivity gains, increases in GDP, and better opportunities on the world market are taken for granted. The communications sector and especially the telecommunications sector are viewed to play a dominant role to reach these ambitious objectives.

The EC commission has become already active with its Green Paper concerning the telecommunications sector which refers to such different things like the terminal equipment market, public procurement, service competition, Open Network Provision, mutual recognition of conformity for terminal equipment and the European-wide cellular radio. Further directives have been announced for the future.

In view of the four characteristic developments of the 1980's and their implications for the communications sector and for communications policy sketched so far the themes of the conference were selected according to the following topics:

(1) Tariff and competition policies,

(2) International and industrial aspects of telecommunications,

(3) Regulatory models and policies,

(4) Convergence of broadcasting and telecommunications.

The papers in this volume have been assigned to these topics and can be found in chapters 2 - 5. In the next section we will give a brief introduction into the central issues and ideas of each paper. As one can see many of the aspects outlined so far have been taken into consideration in detail in the papers.

2. An Introduction to the Studies

Part II of this volume is dealing with different aspects of tariff setting and competition policies. All papers relate to the provision of telecommunications services, however, arguing on the basis of characteristics of different countries. The first paper by Mansell takes an EC country perspective. The actual pricing regimes under network competition in Japan give rise to the theoretical analysis in the paper of Nambu, whilst the third paper by Cave and Trotter examines the situation in the UK.

There are three main issues the **Mansell** paper is dealing with: (1) How can different pricing strategies in telecommunications be characterized and what are the implications of these strategies? (2) What is the telecommunications user's perspective on pricing? (3) What are the trends of rate rebalancing in different European countries since the beginning of the 1980's?

According to Mansell cost adequacy has become more and more important in the discussion of telecommunications tariffs in nearly all Western countries. However, she maintains that there is still substantial room to debate both the appropriate concept of cost and the appropriate cost allocation principles that are applicable.

She sees the need for new pricing structures has been called for mainly by large users. An assessment of the discussion of the EC's Green Paper reveals a clear vote for rate rebalancing of local tariffs and long distance tariffs, for making transparent existing cross-subsidization, and for prohibition of abuse of (monopoly) market power being expressed. Yet, according to Mansell it is necessary to include a wider community of interest in pricing policy discussions in addition to the large business users.

An empirical examination of the patterns of recent structural change in telecommunications tariffs within different EC countries reveals that rate restructuring is underway in the EC. The overall trend suggests that prices for local telephone services are increasing and international tariffs are declining. The pricing of inland services can still be characterized by different national telecommunications priorities. Mansell does not expect a substantial convergence in price structures of switched services before 1992.

Nambu's paper stems from the current Japanese situation of regulated network competition: Except (the former public monopoly now being privatized) NTT up to now three Common Carriers are granted a licence to offer long distance services on the Tokyo-Osaka line. Under the present regulated two-tier price regime the tariffs of the New Common Carriers are lower than NTT's. The New Common Carriers have to pay the local rates of NTT for use of the local loop (which are subsidized by NTT's long distance services), i.e. they actually do not pay an extra access charge thereby participating further in cost coverage.

Nambu presents a theoretical model where the focus is on an optimal rate of entry and an optimal access charge under this competitive situation between an incumbent and several entrants. Optimality refers to maximization of total social welfare in the local market and the long distance market. Total social welfare is measured as the sum of consumer's surplus and producer's surplus. An increasing rate of entry cuts away profits of the incumbent, thus, decreasing the capability to cross-subsidize local tariffs and therefore making it necessary to increase local rates. This increase of local rates is socially optimal in the sense that the rate of entry is determined by the access charge on New Common Carriers.

Cave and Trotter's paper refers to the issues of competition, entry and regulation and their interdependence and characterizes the patterns of the development in the UK since 1984.

Cave and Trotter first review the recent trends regulating basic services, mobile communications and Value Added Services, respectively. As to basic services they focus on the first revision of the price-cap formula, new developments on the resale issue and the envisaged price capping of BT's private circuits. Concerning Mobile Communications there have been different licensing schemes as to cellular phone networks, Telepoint services and private communications networks. As regards Value Added Services Cave and Trotter focus on the difficulties caused by the regulatory dividing line between Basic Services and Value Added Services which is still part of the regulation in this field.

A second focal point of the analysis of Cave and Trotter is the development of the future communication infrastrucuture in the UK and possibilities to launch them. Maintenance of the present duopoly is only one of several options being expressed in the current debate in the UK.

Finally, the paper of Cave and Trotter concentrates on price-capping and entry and discusses different possible levels of regulation ranging from the cental planning solution to a state of minimal regulation where the market forces determine the final outcome. Cave and Trotter do not favor either of these polar cases. Instead, they vote for the partial regulation approach adopted in the UK which is characterized by segmenting markets according to particaular characteristics and thus deriving the scope for regulation.

Part III of this volume covers distinct international and industrial aspects of telecommunications: The paper of Schnöring deals with R&D in telecommunications, the paper of Alleman, Rappoport and Stanley focuses on the current debate on international telephony settlements, and Wieland sheds light on the fundamental role of information and communication technologies and the international division of labor.

In his international comparison of R&D in telecommunications **Schnöring** presents an overview of currently existing national R&D systems, investigates their discrepancies and (dis)advantages and discusses the prospects for future competitiveness both of R&D systems and telecommunications equipment manufacturers.

Schnöring's empirical analysis rests on several statistics of patents and publications and results of interviews. Schnöring first focuses on the serious problems of measurement and definition conducting an international comparison in these fields. His analysis rests on the assumption that basically R&D within a country can be characterized as a national R&D system between four players: the carrier(s), the equipment manufacturers, the state, and universities and other research institutes. He claims that this assumption can be justified for the present situation but would become arguable with a further increase in international co-operation.

Schnöring's paper focuses on the following countries: Japan, France, USA, UK, and West-Germany. He presents the characteristics of their national R&D systems and gives several assessments of the quantitative role of each player within a country.

The paper of **Alleman, Rappoport and Stanley** deals with possible alternatives to the current settlement procedures in international telecommunications services. In particular from less developed countries there is currently a mounting pressure on developed countries to get a larger share from the revenues of international traffic.

The present accounting revenue division between two countries, primarily on a 50 - 50 basis, rests on the accounting rates (normally measured as a rate per traffic unit for a specified service) which become fixed in bilateral negotiations between the countries. Thus, accounting rates virtually always differ from actual tariff rates in a country.

Alleman et al. present a model which incorporates the essential features of a settlement procedure for two countries. Their model takes into account the demand for international calls and the respective cost side for each country, and calculates the net flow of settlements and the net revenue of a country on the basis of the accounting rate and the accounting revenue division. In their empirical analysis Alleman et al. focus on five different scenarios evaluated for a stylized USA vs. Rest of the World case. They present a quantitative assessment of the economic effects of (1) a price reduction in the rest of the world, (2) a change in the accounting rate split from 50 - 50 to 40 - 60 (viewed from the USA), (3) a decrease of the accounting rate, (4) a shift in the demand of the rest of the world, and (5) a decrease in all prices.

Wieland's paper takes on another international aspect and draws the attention to the impact of telematics on an important sector of the world economy, i.e. the banking sector. Wieland's central hypotheses are (1) that without the cost reducing effect of telecommunications the internationalisation in the banking sector could not have occurred and (2) banks in the future will develop to more and more integrated entities, however, specialization will increase. Wieland's argumentation rests on a re-interpretation of Adam Smith's notions of the division of labor and the extent of the market, an empirical appraisal of trends in internationalization of banks and in globalization of financial markets, and an application of arguments of the transaction cost approach to the banking sector.

In **Part IV 4** of this volume regulatory models and policies in the telecommunications field are examined. Olsen concentrates on recent Danish experiences of changing their regulatory policies. Nyevrikel focuses on chances and problems of liberalization of telecommunications in Hungary and Scherer investigates instruments and powers to regulate telecommunications services from an EC perspective.

Olsen's paper reports on how a small member country of the European Community adapts to the changing conditions and expectations brought about by the recent EC policy. Olsen offers an overview of the Danish system of organisation and regulation in telecommunications. According to him the main characteristic of this system in the past is the cooperative element among the different Danish carriers in the fields of use of equipment, traffic planning, prices etc. Regulation in this context has meant primarily the implementation of widely accepted policy lines, reflecting a broad consensus in the Danish society.

Recently, there have been substantial modifications as to the competence of regulation in Denmark. Regulatory functions are now separated from operating functions. Yet, Olsen points out, that the objective of a cooperative solution prevails in Denmark. Olsen discusses recent developments in regulatory policies in Denmark concerning liberalization of terminal equipment, liberalization of telecommunications services and new elements of partial regulation of Danish telecommunications administrations. This analysis leads him to an assessment of the future chances of the Danish system of regulation and the future of Danish telecommunications administrations.

The paper of **Nyevrikel** focuses on a country which has a long tradition of central planning and little experience in market economy which limits in many ways both the kind of admissible changes in the telecommunications field and the speed with which these changes can be accomplished.

Nyevrikel reports that Hungary is now willing to switch to a market economy and she presents the many efforts currently undertaken to establish new institutions and rules e.g. for the banking sector, the tax system and foreign investment. According to Nyevrikel a crucial point concerning the mid-term perspectives of the Hungarian economy is the ownership reform, i.e. the decision about the future of the state owned enterprises.

Concerning the telecommunications sector Nyevrikel presents some rough figures which characterize Hungary presently to be a developing country as to e.g. penetration rates and investment. However, there are fundamental changes underway in Hungary concerning the organisation of the provision of telecommunications services and its regulation. In Hungary network liberalization (on a regional basis) is on the agenda of the political discussion. It can be expected that authority and operational activity will be separated in the near future. Thus, in part there are very close similarities in the current Hungarian debate on the future of telecommunications to that of Western countries.

Scherer's paper focuses on instruments of the Commission of the European Communities to exercise regulatory powers which have already been applied and will be applied in the future in the telecommunications field. Directives have been announced in the EC's Green paper to be the main instrument for legislative regulation of the telecommunications sector. Scherer outlines the requirements to implement these directives on the national levels; he mentions that there is a considerable legal controversy about the direct effects of directives.

One focal point in Scherer's analysis is the discussion of the different legislative options of issuing directives according to Art. 100 and 100 a, respectively, and Art 90 (3) of the EEC Treaty, whereby the first option serves as a basis for a Council directive and the second one is used as a basis for Commission directives.

Another focal point Scherer turns to is the Commission's new approach to standardisation. This approach essentially stipulates that the Council should limit itself to define the objectives and requirements in the general interest by means of directives and should abstain from technical and executive matters. The new approach has shaped already the directive of the mutual recognition of type approval for telecommunications equipment and the ONP directive.

Finally, Scherer draws the attention to the political implications which the recent activities of the EC in the telecommunications law field and the envisaged completion of the internal market until 1992 are likely to have. Scherer does not argue against the many advantages brought about by these developments, however, he claims that there also might be political costs. These costs e.g. can be seen in a further deparliamentarisation of telecommunications policy-making and in the development of a new administrative network between EC and national telecommunications authorities.

Part V of this volume focuses on issues around the convergence of telecommunications and broadcasting. The paper of Lera turns to possible scenarios for integrated broadband communications without special reference to a particular country, whereas the paper of Hanada and that of Delcros and Chamoux concentrate on distinct national aspects of convergence in Japan and France, respectively.

Lera's analysis starts with the statement that both from an economic and a technical viewpoint there is a rationale for an integrated broadband infrastructure capable of providing classical and future telecommunications services (e.g. image telephony) as well as broadcasting services of today and tomorrow. However, Lera points out, that the dynamics in the developement of these two fields do not lead automatically to a convergence. Thus, the focus of his paper is on one of the several possible limiting factors which according to him lies in the respective regulation schemes.

Lera's analysis first concentrates on the current organisational patterns of the two systems and the respective operating characteristics. He then examines the current status of several regulatory aspects for each of the two systems. Among these aspects are the legal character of the services, the role of authority and the operating branch, the access to the network for third parties, the network integrity and standardisation, and price regulation. Lera's viewpoint is that there are substantial differences between regulation of telecommunications and broadcasting services as to these aspects. He then concludes that there are at least two possible scenarios for a future broadband infrastructure: (1) Planning and operation of this infrastructure is the prerogative of a single service provider and (2) there is network competition.

Hanada's paper takes for granted from the outset that there are processes leading to a convergence of economic and institutional aspects concerning broadcasting, telecommunications, and information processing, however in some respect there may also be a divergence of certain aspects. He claims that convergence takes a different form in different countries depending upon legal and political factors as well as historical experiences in each country.

His paper turns to the Japanese situation whereby his analysis focuses on the present legal and regulatory framework of broadcasting and telecommunications, the driving forces behind convergence, recent political issues discussed in the context of convergence, the functional definition of broadcasting and telecommunication and implications for the process of convergence, and industry specific and regulatory consequences of convergence.

According to Hanada the characteristic of the present Japanese broadcasting system is the "unification model" of the broadcasting enterprises responsibility both for the network and the information content. In the field of telecommunications there recently have been major steps to liberalisation, in particular concerning network competition. Presently, no administrative regulation is applied to the contents to be transmitted by telecommunications, while in broadcasting there are various regulatory constraints.

Hanada views primarily technical progress, an increasing diversification of communication needs, liberalised telecommunications environment, and the vision of the Japanese Ministry of Post and Telecommunications of future Integrated Services Digital Broadcasting networks as factors promoting convergence in Japan.

Hanada reports on vivid political discussions and decisions in Japan relating to e.g. cable networks providing telecommunication services, telecommunications satellites and broadcasting services and vice versa, the dominant telecommunications carrier NTT's wish to construct a cable television network and the beginning of its provision of broadcasting service via its telephone network.

Concerning functional differences between telecommunication and broadcasting Hanada distinguishes three dimensions: (1) the mode of supplying information, (2) the access qualification for information, and (3) the information processing functions. Hanada concludes that the technological convergence of broadcasting and telecommunications is giving rise to structural changes in the organisations and markets related to broadcasting and also brings about changes in the relationship among various industrial sectors in Japan. There will be a need for a new concept of broadcasting and a new regulatory framework based on it. Hanada furthermore claims that convergence generates an increasing responsibility of policy - making in particular with respect to cultural and social values.

In the paper of **Chamoux and Delcros** the focus is on telecommunication and audiovisual legal regimes in France. Chamoux and Delcros state that the passing on of the 1982 Audivisual Law abolishing the monopoly given to Télédiffusion de France was the beginning of a distinct legal activity in this field. Their analysis focuses on the relationship between this law and the telecommunications law and the recent developments of licensing and licences in France for the provision of these two service categories.

Chamoux and Delcros report that the responsible licensing bodies (for audiovisual and broadcast as well as for telecommunication) as well as their competences often have changed in the past and that these changes have been dependent on the respective political power in charge. It was only recently that the audiovisual regulators gained full power to license any facility and service in their media, in particular to regulate and allocate those parts of the spectrum allocated to radio, TV etc. Chamoux and Delcros report on the current discussion in France whether or not frequencies can be appropriated (i.e. being a real asset).

As to network licenses for telecommunications a draft legislation is announced for 1990 which according to Chamoux and Delcros may introduce a limited monopoly constrained by the conditions of the EC's Green Paper.

Chamoux and Delcros state that a period has ended where a convergence of audiovisual and telecommunications was likely to be expected and that France now is bound to a diverging set of rules concerning principles and practical terms and that there is also a divergence with respect to the regulatory body in charge of these rules.They view mobile radio and Cable TV to be the two main issues in the near future which both may force legislation in France to realise legal practice in telecommunications and audiovisual operations.

Part II

Tariff and Competition Policies

2

Restructuring Telecommunication Tariffs: Policy Issues, Trends and Implications

Robin Mansell[1]

Centre for Information and Communication Technologies, Science Policy Research Unit University of Sussex Brighton

1. Introduction

With the release of the now familiar 'Green Paper on Telecommunications' in 1987, the Commission of the European Communities (CEC) initiated a discussion on the future of telecommunications development within the European Community (EC). The policy positions described in this document stimulated debate on the complex problem of ensuring that European telecommunication network and service infrastructure will support the needs of the Single European Market in the 1990s and beyond. Policy, regulatory and structural issues were addressed. By late 1989 many of the policy positions that were advocated in the 'Green Paper' were in the process of being

---

1 The research for this paper was supported in part by the OECD Information, Computers and Communication Policy Division and by the UK Programme on Information and Communication Technologies. The author is grateful for comments by Dr. Kevin Morgan. Any errors and omissions are the author's responsibility.

implemented through Council Directives on equipment procurement, terminal equipment deregulation and service liberalisation. Work continues in a host of other areas including standards, interconnection of private and public networks and conditions for network access and use, and tariff policy.

On the issue of telecommunication service tariffs, the CEC has called for movement toward prices which reflect overall cost trends. A process of consensus-building as to the interpretation that should be given to 'cost-oriented' tariffs is underway. But even in the unlikely event that consensus is reached by 1992, substantial differences in tariff structures within the EC are likely to remain. The CEC considers that "telecommunications is the most critical area for influencing the 'nervous system' of modern society"[2], and it's attention to telecommunication tariffs reflects its concern with the implications of discrepancies in pricing policies and price structures. Tariffs are among the main tools that can be used to influence the evolution of the telecommunication infrastructure. They affect the rate and type of investment in public and private networks, the growth of traffic within and among countries, and they create incentives which can stimulate or dampen the growth of telecommunication network-based services of all kinds.

The CEC has called for:

> "..recognition of the fact that telecommunication tariffs should follow overall cost trends and that a certain amount of rebalancing of tariffs will be inevitable, as far as compatible with public service goals... a fair trade-off between cost-orientation and the aim of universal service on reasonably the same terms for all will have to be developed".[3]

In its 1988 paper on the implementation of the positions outlined in the earlier 'Green Paper', the CEC's attitude toward tariffs is considerably more forceful: "If reasonable overall cost- related tariffs are not achieved by 1st January 1992, the whole approach with regard to the future evolution of the telecommunications sector will have to be reevaluated".[4] Disparate tariff policies and practices among the Member States are unlikely to provide the 'level playing field' envisaged by the concept of the Single European Market. Prices for services that bear little relationship to costs, and which are

---

2 Commission of the European Communities (1987) "Towards a Dynamic European Economy: Green Paper on the Development of the Common Market for Telecommunication Services and Equipment." (COM(87)290 final, June, p. i.

3 Op. Cit., pp. 14, 82.

4 Commission of the European Communities (1988) "Implementing the Green Paper on the Development of the Market for Telecommunications Services and Equipment." (COM(88) final, February, p. 18.

established to achieve a variety of economic and social policy objectives, can create barriers to the flow and use of information within the EC.

As well as enabling Member States to maintain or erect non-tariff barriers, telecommunication price structures also have distributional implications. The CEC is aware of the need to ensure that price restructuring does not unduly disadvantage some groups of service users. And, although prices are not the only factor that influences the development of service applications, they are central to the capacity of the telecommunication infrastructure to stimulate economic activity.

There are several widely acknowledged factors that have created the need for substantial rethinking of traditional telecommunication tariff setting practices. The impetus toward tariff restructuring stems from: technical change and shifts in underlying telecommunication facility cost structures; increasing competition; the perceived threat of bypass and revenue erosion due to simple resale; the alignment of tariffs with those in other countries; and attempts to reduce the cost of telecommunications for the business community.[5]

Tariff restructuring is undoubtedly inevitable in the sense that public telecommunication operators (PTOs) must shift from the use of administered monopoly prices to accommodate a more competitive service environment. Restructuring is needed to ensure that PTOs do not lose business simply because their rate structures have not been related to costs. However, new price structures cannot be introduced without considering their impact on different types of service users and historical pricing practices. Although there is widespread awareness that tariffs for public switched telephone network (PSTN) and leased line services are being restructured, most policy discussion has focused on comparisons of the relative level of prices among countries in the EC. However, telecommunication pricing is a matter of national concern and the rate at which countries engage in radical price restructuring is likely to reflect the overall direction and magnitude of changes that have been occurring over the past decade within countries and the extent to which these are in line with Europe-wide dominant trends.

The rationale underlying the process of tariff restructuring is rarely discussed except from the self-interested perspective of suppliers and powerful large user lobbying groups. Pricing issues are also often debated within the framework of the theoretical world of optimality and 'second-best' solutions which take little account of the social

[5] OECD, Telecommunications Network-based Services: Policy Implicaitons, ICCP Report No. 18, Paris 1989

and economic factors that decision-makers must consider in moving toward new price structures. As a result, debate in Europe generally has failed to consider the full range of alternative price strategies and options that could be implemented. An air of inevitability tends to suggest that there are few, if any, variations in pricing strategies that could be considered if the CEC's policy objectives are to be met.

Thus, the predominant view can be summarised as follows. Technical change and competition cause prices to move closer to cost. Costs are declining unevenly among different parts of the network infrastructure. Substantial price increases are warranted for local telephone service and long distance telephone prices should decline. These changes in the traditional structure of tariffs provide customers with more accurate price signals and stimulate demand -- or so the argument goes. Such prescriptions lead to sporadic outcries from consumer groups who worry about the fate of residential customers who may no longer be able to afford local telephone service. Their arguments are usually not based on a careful consideration of the combination of factors that have led to price changes because of the cloak of mystery that surrounds pricing debates, i.e. the relationship between prices and 'costs' is not considered, and there is a general absence of data in the public domain.

This paper offers an analysis of several major issues in telecommunication pricing. It examines some of the ways in which these issues are viewed by different communities of interest and considers the implications of alternative pricing options within the scope of trends toward rate restructuring, (i.e. reductions in long-distance telephone charges, and realignment of prices for switched and unswitched network-based services). An analysis of patterns of price changes within European countries is used to illustrate the diversity of trends which have appeared from 1980 to 1987.

2. The Cost/Price Controversy

Telecommunication tariff restructuring is both cause and consequence of technical, structural and regulatory change. It is influenced by the diffusion and application of existing and new services and by the gradual introduction of more competitive markets. With the development of new technologies and alternatives to PTO service supply, pressures have mounted to bring service prices in line with costs on the grounds of economic efficiency. For many PTOs, tariff restructuring is needed to compensate for traditional subsidies. Many types of subsidy underlie existing rate structures, but one of the more visible and controversial is the relationship between the costs and prices associated with long distance and local telephone service. For example, "financial transfers from long-distance services can be approximately estimated to correspond to up to 25% of the overall telecommunications revenue of more than 60,000 million ECU in the Community, i.e. roughly 15,000 million ECU"[6]. Such subsidies lie at the heart of tariff restructuring controversies and they are crucial to whether the objectives of 'universal' or 'public' telephone and other telecommunication services will be maintained and extended in the future.

Another source of pressure toward tariff restructuring is the expansion of service offerings beyond voice telephony. Many large firms and multinational companies are establishing their own private networks to achieve greater control and flexibility over their business communication needs. Traditionally, PTO's have supplied voice telephone service and a limited number of other services. Today, there is an increasing variety of telecommunication network-based services (e.g. transaction, information and communication) that can be supplied over the common public network.[7] Since PTOs are unlikely to be able to meet all demand, tariff structures must be redesigned to accommodate new service suppliers and to allow for expanded consumer choice based on a rational and efficient tariff structure. In this sense, tariff restructuring reflects a shift from supply to demand-based services and price structures. PTO tariff structures of the future will have to accommodate existing and new services, new regulatory and trade frameworks and competition in some markets. In addition, public networks supporting a limited number of voice grade services are being superseded by common public networks that are used to supply a wide variety of very different types of services.

Although they are not mutually exclusive, there are two 'world views' on the costing philosophies that underlie telecommunication price structures. The first, known as 'cost-of- service' pricing reflects the perception that price structures should lead to an optimal allocation of resources, as would occur in competitive markets. This approach focuses on maximising economic efficiency, stimulating investment and the general movement of prices to accommodate the introduction of new telecommunication infrastructures and competition. The second, 'value of service' pricing, is associated with traditional tariff-setting practice, national industrial and economic policy, social and public service objectives and a monopolistic telecommunication framework.

---

6 H. Ungerer and N. P. Costello. Telecommunications in Europe: Free Choice for the User in Europe's 1992 Market, European Perspectives Series, Brussels, 1988, p. 44

7 Transaction services include services such as teleshopping, telebanking; information services allow users to access information held in remote databases; communication services include electronic mail, computer conferencing, etc.

Despite the currency of the former in today's debates on tariff restructuring, both 'cost-of-service' and 'value of service' pricing find their roots in the history of telephony. Differences of opinion on the role these concepts should play in telecommunications are confounded in practice by priorities accorded to different aspects of policy. A wide gulf exists between telecommunication price theory, understanding of telecommunication cost/price relationships, and empirical evidence.

Tariff restructuring in support of economic efficiency can be characterised as focusing on the need to achieve economic efficiency in the underlying telecommunication system. An efficient infrastructure is fundamental to developing the market for goods and services. Universal or 'public' service policies are often juxtaposed with these efficiency goals. Consideration is given to maintaining the value of the telephone network (network externalities) and protecting subscribers where these objectives conflict with efficiency objectives. Given the formal and public nature of telecommunication regulation in the United States and Canada, it is not surprising that 'cost of service' and 'value of service' pricing have been fully discussed. In contrast, in Europe, there is general agreement that telecommunication tariffs historically have not been designed on the basis of costs. Under monopoly supply, this permitted the pursuit of a variety of objectives without significant pressure from the market place. For many PTOs pricing policy included subsidising certain types of subscribers, the postal service or general government revenues. Detailed analyses of costs of services and their relationship to prices has generally not been undertaken on a regular basis, and the accounting and statistical records needed to measure costs have not, or have only recently, been developed.[8]

When the experience of most European PTOs is examined in the light of debates in the economics literature on pricing, it seems that there was no conscious decision to accept or reject cost as a basis for establishing tariff structures. 'Cost' was not the criterion that was used. Rather tariffs were set in response to more global objectives. Most PTOs did not feel it necessary to attempt to determine which services were being subsidised by what other services until the advent of competitive pressures.

The experience in North America has been different. There have been ongoing policy debates and numerous studies examining pricing policies and the role of costs in telecommunication tariff structures. These appear in the professional (economics, law, and policy) literature, in decisions by regulatory authorities and in the United States, in

8 See T. Johanson, D. Lewing, and D. Rogerson (1989). "Cost based Tariffs for Data Services: Positions and Trends in Europe." London: Ovum Ltd. PTOs in France, the FRG, Italy, Spain, Sweden and the UK have developed cost accounting systems which allow them in some cases to calculate the costs of categories of service.

court decisions. These debates about the criteria for designing tariff structures have a long history in other industries with public service characteristics, most notably transport and electricity. The appropriate structuring of rail and electricity tariffs spawned serious debate and in- depth study before the turn of the century, raising virtually all of the issues now being raised in telecommunications.[9]

The economic theory of pricing indicates that market prices are determined by the conditions of supply and demand. The quantity supplied depends upon the cost of production. Higher prices will result in more supply. The quantity demanded depends upon the 'value of the service' to the purchasers. Lower prices will elicit greater demand. Historically, the pricing debate has been about the appropriate role of 'cost of service' and 'value of service' principles in the design of pricing structures, the appropriate concepts of 'cost' and 'value' that should be used, and the best methods of measuring these concepts.[10]

In the economic theory of competitive markets, prices are driven to costs (including a fair return to cover the cost of attracting capital). The amount of demand that is met is for those who will pay the cost-based price. All those buyers who are willing to pay a price that covers the cost of production will be served on a non-discriminatory basis. The analysis of the competitive market model has led some parties to the conclusion that telecommunication service prices should be based on 'cost of service'.

Others argue that because there are substantial economies of scale and scope in the supply of telecommunication systems and services, the marginal (or incremental) cost of adding higher volumes of service and new services will be less than the average unit cost. They claim that the appropriate cost concept is marginal cost, which should represent the minimum price. Actual prices should be set at marginal cost or above, depending on 'value of service' (demand elasticity) considerations. Thus, the primary determinant of the structure and level of prices for different services will be 'value of service'.

9 See J. C. Bonbright (1962), Principles of Public Utility Rates. New York: Columbia University Press and H. M. Trebing (ed.) Diversification, Deregulation and Increased Uncertainty in the Public Utilities Industries. East Lansing: Michigan.

10 It should be recognised that there is a different concept of "value of service" that can be used in telecommunications. "Value of service" sometimes refers to the number of telephones that can be reached within a local area from a local telephone. Thus, there is greater "value of service" in a community of 100.000 telephones than in one of 10.000. This is sometimes used as a basis for charging higher rental charges in the community with greater "value of service". In economic terms, this is a difference in quality of service. It is an entirely separate issue from the treatment of "value of service" as a measure of intensity of demand at alternative possible price levels.

Proponents of 'value of service' pricing have not proposed unbridled monopolistic pricing. There is some agreement that such a pricing practice would amount to exploitative and discriminatory monopoly pricing. Instead, 'value of service' pricing should only be used to permit the telecommunication monopoly to cover its total cost (or revenue requirement) including a fair return on its invested capital. In the earlier debates on railroad and electricity pricing, this was commonly referred to as using 'value of service' pricing to cover the overhead or common costs. According to Clark:

> "overhead costs must be levied upon those parts of the business as will stand the burden, while other parts of the business, which otherwise cannot be had at all, are charged what they can pay, regardless of overhead costs"[11].

In the terminology of the more recent literature in electricity and telecommunications, this is referred to as applying a mark-up of price over marginal cost in terms of relative inelasticity of demand (e.g. Ramsey pricing). The most inelastic demand receives the greatest mark-up of price over marginal cost. For example, local telephone service is believed to have a relatively inelastic demand compared with long distance service. Therefore, it is argued that local service should have a higher mark-up over marginal cost than long distance service.

Most PTOs do not currently have the necessary information to implement any version of either side of the traditional telecommunication debate on price structures. New accounting and information systems are being introduced. These systems are being developed in response to the implementation of multiple services on a common telecommunication system. But such systems will not resolve cost and pricing controversies. It is worth emphasising that cost is part of all approaches to price structure debates. In the new telecommunication environment PTOs may elect to use cost concepts differently in achieving their policy objectives. But for telecommunications network suppliers and service users, greater attention to cost is not only inescapable, it is essential to the successful implementation of new telecommunication policies.

Even when it is acknowledged that cost must play a more significant role in telecommunication pricing, there is still substantial room for debate over the appropriate concept of cost that is applicable. The treatment of common and overhead costs associated with network supply is particularly controversial. Many 'cost of service' advocates argue that most, if not all common and overhead costs can be allocated on a cost causation basis. In contrast, 'value of service' advocates suggest that common and overhead costs cannot be allocated on a cost causation basis. Instead, the additional

11 J. M. Clark (1923), The Economics of Overhead Costs. Chicago: University of Chicago Press.

revenues necessary to cover the total costs should be realised by charging prices in excess of marginal cost, thus reflecting 'value of service' (i.e. the relative demand elasticity) considerations.

In summary, one approach proposes to recover common and overhead costs by allocation to services on the best estimates of cost causation. The second approach proposes to recover common and overhead costs by means of differentiated mark-ups above the marginal or direct costs of particular services. Unfortunately, in much of the debate on 'cost-based' pricing, the distinctions between 'cost of service' and 'value of service' pricing, and their implications, are not clearly presented.

This facet of the debate has been pursued in greatest depth in the United States. 'Cost-based' pricing, at first glance, is more attractive than pricing 'without regard to cost'. Costs can appear to have a good deal more quantitative rigour and empirical criteria can be devised as the basis for tariff-setting. But unanimity on what the criteria or cost standards should be is very difficult to achieve.

Analysis and debate on cost standards generally centre on the common cost problem.

> "At the core of almost all the pricing issues in telecommunications is the fact that the products of this industry are a large and increasing diversity of services issuing from common facilities...the issues tend to be framed in terms of the proper apportionment of their common costs among them, a task further complicated by the fact that some of the services have become subject to competition, while others continue to be offered by a single supplier, under close regulation".[12]

It can be assumed, on the one hand, in the determination of telecommunication service costs, that a major share of costs associated with the local exchange network do not vary with usage. These costs should therefore be recovered from charges to the local telephone service subscriber regardless of actual use and common costs should not be apportioned among the various services that use these common facilities. Since traditional price structures generally have set prices for long-distance service above its direct costs, the argument is made that local telephone service is being subsidised by long distance service. Tariff restructuring should occur to bring prices into alignment with 'costs'.

On the other hand, there are those who argue that to judge the reasonableness of prices, it is necessary to allocate common costs in proportion to their relative use by different

12 A. E. Kahn and W. B. Shew (1987), "Current Issues in Telecommunications Regulation: Pricing," Yale Journal of Regulation, Vol. 4, No. 2, Spring. p 194

services. In this case a proportion of the costs associated with commonly used exchange plant and equipment is allocated to services in relation to how the services benefit from the use of common facilities. Using this approach, prices for services that depend on a common network reflect the design characteristics and associated costs each service incurs in its production. Cost studies using this approach generally show that traditional prices for long distance service make a reasonable contribution to common costs.

Even from this second perspective, there could be justification for rate restructuring on grounds other than cost. For example, there may be reason to recover substantially greater revenues from local telephone service subscribers to reduce costs to business users who rely more heavily on long distance service. New pricing strategies could be designed to stimulate the diffusion of new services or to implement other industrial policy objectives. Changing technologies also provide a basis for tariff restructuring, but the common cost problem remains.

There are fundamental differences in approaches to pricing between the United States (and to some extent Canada) and countries in Europe. In the former case, regulation has required that prices be set in reference to cost. The question has been one of the appropriate cost standard and the grounds for departures from it. In Europe, PTOs generally have been left to their own devices in setting prices. 'Public service' responsibilities have provided the constraints within which prices have been set.[13]

This situation is changing rapidly. There is growing agreement in the EC on the principle that service prices should bear some relationship to production costs. Disagreement surrounds the measurement of cost and the circumstances under which prices should deviate from cost to implement policy objectives (i.e., social policies to extend access services, economic policies to stimulate demand, etc.). Despite the fact that European PTOs are moving rapidly (or gradually) toward cost-related pricing, prices invariably will depart from cost. Broader political and economic considerations mean that even in the presence of greater cost detail and consensus on cost/price relationships, such information can only provide targets for PTOs. Detailed pricing decisions are best left to PTOs who are well-placed to analyse changing market demand and to propose the price structures that will meet various economic and social policy objectives. The results of subsequent tariff restructuring can then be monitored by policy-makers.

[13] N. Garnham (1988) "Universal Service in European Telecommunications," OECD, Paris, June.

3. The Telecommunication User's Perspective

Since 1987, the CEC has called for submissions on its proposals for developing the new environment for telecommunications. Although the telecommunication user community is extremely heterogeneous it is mainly large business users that have actively participated in this process. A sample of some of their comments shows the extent to which they support movement toward new pricing structures that reduce their overall telecommunication costs, provide greater stability in tariffs and generally move in the direction of 'cost-related' pricing.[14]

The American Chamber of Commerce in Belgium, for example, "endorses the Commission's intention to promote a cost-oriented approach for telecommunications tariffs for reserved [e.g. voice telephone] services" and "shares the view of an inevitable trend towards rebalancing local and long distance tariffs". They argue that there is no inevitable conflict between 'universal service' obligations and cost-oriented pricing. Themes that have been debated in the United States resonate in the Chamber's view on the benefits of cost-oriented tariffs:

> "An increase in local rates and a decrease in long distance rates - while this will have an immediate positive effect on the business community, it will also be beneficial for the whole of society as it will help offset the inequities between residents of large urban areas and those in remote regions. If there is a social need to subsidize local rates, it should be done in a 'transparent' way either from public funds or from revenues of the same service (i.e. telephone service). We see it as unfair to cross-subsidize among different types of services (i.e. telephone service subsidized from public data network service or vice-versa)".

The Council of Netherlands Industrial Federations says that: "the tariff structure for each transmission service of a PTT should be consistent with the network-facilities used, and be based on the costs of the service provided...a system of 'charge what the market can bear' is utterly out of place in a monopolistic environment".

The European Computing Services Association "agrees that the setting of tariffs is a major component of a Telecommunications Administrations' overall commercial strategy; but the abuse of dominant positions must be halted. Increasingly pressure must be exerted for local tariffs to follow true basic cost trends. As the cost of provision of long distance traffic has fallen much more than the cost of provision of local traffic, tariffs should become less distant-dependent".

---

14 Commission of the European Communities (1988) "Comments on the Green Paper up to February 1988," Brussels, XIII/41(88)

IBM Europe's position echoes that of other large telecommunication users: "We observe that there is a critical need to rebalance today's distorted tariffs for switched services, not only as a matter of equity to business users but also as the best means to assuage Administration fears of 'cream skimming'". IBM argue that reserved services should be priced on the basis of costs (including operating expenses, interest on loans, depreciation and a reasonable return on investment).

Business users see a need for a major change in the emphasis given to economic and social policy objectives. The International Chamber of Commerce notes that "...telecommunication networks can no longer be considered merely a convenience. Business relies on their effective and efficient operation in order to develop in the same way as raw materials are required for the manufacture of goods. Companies will invest in an area where they can get facilities and plan for costs, but the reverse is also true".

Although discussion on pricing issues tends to reflect large volume user views, there is controversy over telecommunication pricing in the political realm. Some countries want to keep local service rates at a low level to implement social, regional and other policy objectives. Consumer representatives (e.g. Bureau European des Unions de Consommateurs) have expressed concern about the combined impact of technological change, a more commercial orientation on the part of PTOs, and rate restructuring. For example,

> "Partly because there is likely to be more competition in the business sector of telecommunications and partly because business users are getting organised to negotiate with telecommunications administrations, telephone charges to residential consumers may well be looked on as a source of monopoly profits, to be used to subsidise other services".[15]

In summary, the themes among large firms and their representative organisations are consistently in favour of rate restructuring.

By any definition of 'cost', if 'cost-related' tariffs are being implemented it seems reasonable to expect some public body to know what costs and cost standards are being used. If this is not the case, it is impossible to determine whether tariffs are cost-related or whether they unfairly disadvantage certain types of service users.

The CEC has made clear that it does not intend to establish systematic monitoring of prices (or costs) since tariff-setting is a component of PTO commercial strategy. However, tariffs for telephony and for other services (in the context of Open Network

15 BEUC (1988) "Consumers and the Telephone Services in the EEC". BEUC/238/87, July, Brussels, p. 3

Provision) are being examined. The question as to how 'cost' is to be determined is receiving attention within the Commission and PTOs will be expected to be able to demonstrate the general relationship between costs and tariffs. The CEC's Competition Policy Directorate (DG-IV) is not interested in the level of tariffs except where there may be instances of possible abuse of dominant position under the rules of the Treaty of Rome. The Commission's real interest is in the structure of tariffs and the principles used by PTOs and these are expected to become more transparent.

4. Selected Trends in Switched Service Pricing

Telecommunication prices are controlled indirectly by rate-of- return regulation (United States and Canada); by 'price caps' (United Kingdom, Australia and proposed in the United States); and by less formal governmental review processes (other countries). Users make decisions about the mix of network-based services they will use in the light of many considerations, but the level and structure of prices are clearly important.

As has been indicated in Section 2, tariff restructuring is attributable to many factors. Apart from increasing competition, one of the most significant factors is the introduction of new network capabilities derived from innovative technical configurations. Commercial Integrated Services Digital Networks (ISDN) and the increasing availability of virtual private networks have created pressures to introduce rate restructuring as PTOs introduce new price relationships between their switched and unswitched service offerings. If there is an eventual diffusion of ISDN to most customer premises, the cost of accessing telecommunication networks could rise significantly. For the foreseeable future, the Public Switched Telephone Network (PSTN) will be an option for subscribers, but it could become necessary for PTOs to raise PSTN prices in an effort to migrate users to the ISDN.

A variety of new tariff offerings have been introduced in the United States and in some countries in Europe. These tariffs apply to new networks, such as AT&T's Software Defined Network (SDN), MCI's Vnet and Sprint's Virtual Private Network (VPN) and they provide different types of connections and access to transmission facilities. The charges for these offerings and other switched network services (e.g. voice telephone and leased circuits) are creating a complex set of tariffs for the large user. As services become increasingly competitive, the pressure to reduce transmission costs is likely to grow. The combination of reductions in long distance prices, could result in escalating local service prices.

Other pressures on rate structures stem from the increasing costs of software which supports virtual private networks. These software costs may be reflected in charges for local exchange voice service since software design features and their related costs are difficult to desegregate among services that depend on the common public network infrastructure. Thus the treatment of common network costs for pricing purposes could become substantially more complex in the all-digital environment of the future.

Some aspects of the rate restructuring problem differ in magnitude as between the United States and Europe. In the United States, for example, a package price is often charged for access to the telecommunication network and for local service calls. In other countries the practice is to charge separately for access and usage of the local telephone network. In the United States, allocation of common network costs associated with the local exchange must be made to local and to long distance services. Given that such costs often comprise 50 to 60 per cent of the investment in the typical telephone network this is a major task. In contrast, in Europe the problem is one of determining the appropriate costs of connection, rental and usage. However, this difference does not eliminate the problem. Judgements are still required on the components of cost that should be recovered by 'local' usage tariffs and by connection and other charges. The discussion of pricing philosophies in the preceding section (Section 2) has highlighted the theoretical debates on the question of tariff principles that could be adopted by PTOs. In the pragmatic world where PTOs must establish tariffs, they are confronted with the fact that price elasticities are not known with any degree of precision and measures are subject to different interpretations in monopoly and competitive environments. They also have to contend with the fact that cost allocation procedures are always imperfect. In addition, the theory underlying alternative pricing scenarios has some deficiencies. In particular, 'value of service' (or Ramsey pricing) does not provide easy solutions in cases where oligopolistic conditions prevail since price elasticities tend to become relatively meaningless as one moves away from fully competitive market conditions. The theory also fails to deal adequately with the fact that network local access is a necessary condition for the use of many services. Thus a prescription that suggests that prices for subscription should increase may cause revenue losses.

Regardless of whether 'cost of service' or 'value of service' is advocated, the general expectation throughout the European Community is that prices for access and local calls should rise as quickly as possible and that prices for long distance and international calls should be reduced. Although most countries support this general principle, there are substantial differences in the extent to which they are implementing the necessary tariff restructuring policies. There is increasing discussion of pricing issues, but so far very little empirical data on past practices has been available. Even less evidence on the actual price changes that have taken place over time has been provided by PTOs in most EC countries. This section of the paper takes a step toward clarifying the patterns of structural change in telecommunication tariffs for Public Switched Telephone Network (PSTN) services within EC countries.[16]

4.1 Methodology

Patterns in tariffs for selected services are examined over the 1980-1987 period. The analysis does not consider cross-country comparisons of relative tariff levels. The objective is limited to examination of general trends in tariffs over time using a consistent data source and method of calculation.

Any database collected over time and covering a large number of countries must be expected to contain some discrepancies as compared to nationally collected databases. Such deficiencies can account for differences in results in some cases. The purpose of the present exercise is to provide an overview of overall structural trends in rate restructuring. It is not intended to offer prescriptions for pricing policy in the future.

A consistent data set containing charges for services was needed that met criteria such as reasonable coverage of EC countries; consistency in reporting tariffs over the 1980/87 time frame; and reporting of elements of service costs, e.g., local, trunk, international and by distance. These criteria were met by the Tarifica database which was provided by Logica UK. This data source reports charges for services on a cost per minute/cost per month basis. In the case of cost per minute usage charges, these data cannot be used to determine costs for multiple minutes of use. They also cannot be combined with usage data to develop profiles of charges for the 'typical user'.

Given the limited resources available for this analysis, certain parameters that can affect the impact of price changes on the cost of service to users have not been addressed. Thus, for example, regional variations within countries in price structures and intercity communication costs have not been examined. Variables such as the average length of call have also not been factored into the analysis. Despite these limitations, the results provide an overview of the types of general movements in prices for different components of telephone service charges over a substantially longer period and at a greater level of detail than is available elsewhere.

---

[16] The analysis draws on a longer report by R. Mansell, "Telecommunication Rate Restructuring: Issues, Patterns and Policy Problems for OECD Countries", prepared for the OECD/ICCP/TISP, November 1988.

The results reported here focus on patterns of changes within each country for telephone service tariff elements. The main purpose is to examine the general direction and magnitude of rate restructuring over the period. It must be recognised that the calculation of percentage changes from 1980 to 1987 is sensitive to prices in effect in the base year and end year. By choosing 1980 as the base year, the preceding history of tariff changes is not reflected in the results.

In addition, some observed changes are a reflection of the fact that countries are starting from different positions in the base year and the results cannot be attributed simply to tariff restructuring. However, as the analysis below suggests, there are sufficient common trends in the expected direction to confirm that the methodology can serve as a proxy until more detailed studies are undertaken.

The Tarifica data are reported in national currencies (current prices) in each update in a country series. It is possible to convert these data using OECD exchange rates/PPPs, but this has not been done to discourage the temptation to use these results to examine changes in the relative level of prices among countries.

The determination of service charges for each of the years 1980/87 relied on several 'rules of thumb'. All tariff updates from November 1979 to January 1988 for each element of service and country were examined to determine the charge in effect for the longest period within any given year (data used are January 1980 to December 1987). For example, in some cases the data reflect tariffs in effect for a full year (or more), in others for only several months. This method has the advantage of avoiding the need to average different price levels within a year. It has the disadvantage that it fails to reflect significant changes within a year. However, a check on the data has revealed that most changes occurring within a year have been captured in the preceding or following year. In cases where anomalies in reported data have been found, they have been checked with preceding or subsequent issues of the Tarifica data. Data on discounts (except for time of day) and/or multipliers applied to charges for use of the same service for different applications have not been included.

In order to investigate variations among peak and off-peak periods and tariffs for usage over shorter and longer distances, it has been necessary to specify the rate elements to be used for each country and service tariff element. This has involved a degree of judgement.[17]

17 Because this analysis is concerned with the overall magnitude and direction of changes in tariffs, details on the specific rate elements used are not provided here. It is therefore, not possible to recalculate the results using an alternative data set from the information in this paper.

Charges for services are reported in current and constant (1980) prices. The use of current prices enables an examination of the magnitude and direction of changes in tariffs that have been introduced. These changes can be attributed to the actual changes in tariffs which have been implemented by PTOs. Results in constant prices indicate the impact of tariff restructuring in terms of real money values, but this method of reporting camouflages the changes in tariffs that have resulted from the implementation of policy decisions that are based on many factors in addition to the desirability of ensuring that prices for services do not exceed the rate of inflation.

4.2 Restructuring PSTN Tariffs

Tables 1 and 2 illustrate changes in the structural relationship among tariff elements for telephone service between 1980 and 1987 for the EC countries. In Table 1 where current prices are used, it is possible to examine the magnitude and direction of changes in tariff elements for service components and to attribute these to changes in policy. Where tariffs are shown to have increased, a portion of the increase is attributable to inflation. Therefore, the interesting comparison within each country is the relationship among prices for components which have increased, remained stable or decreased over the period. Each country represents very different conditions for the development of telecommunication infrastructure. Pricing policies should be expected to vary as a function, for example, of population, geographic size and topography, telephone penetration, pace of network modernisation, and priority given to 'public' serivce goals and to industrial policy goals.

Table 1 shows percentage changes (calculated using current prices) in the various components of charges for telephone service. A pattern of decreasing prices for international telephone service is visible with the exception of Greece, Ireland, Italy, Portugal, and Spain (effects of price changes are hidden by high levels of inflation over the period in these countries). The trends toward lower international tariffs can be contrasted with the increases for exchange line installation and rental charges and for some types of inland calls, especially those over shorter distances. Exceptions to these trends in the case of Belgium, France, the FRG, and Spain suggest the range of options that are available within the general framework of tariff restructuring. For example, France has a policy of reducing installation and rental prices, coupled with increases in the prices for inland calls.

Table 1: Patterns of Tariff Restructuring Telephone Service, Percentage Changes 1980/87, Current Prices

Country	Exchange Line		Inland Calls/Min.[1]						Internat. Dialled[2] Telephone Calls/Min. Range of Price Incr/Decr.[3] (From Country to 18 countries)	
	Installation charges	Rental charges	Peak			Off Peak				
			Loc	Short Distance	Long Distance	Loc	Short Distance	Long Distance	High	Low
Belgium	-11	9	N/A[4]	20	20	N/A	20	20	22	-17
Denmark	39	68	N/A	57	13	N/A	48	12	9	-28
France[5]	-64	-17	46	46	35	3	3	-3	9	-41
Federal Republic of Germany	-68	0	1	0	0	-1	0	74	0	-50
Greece	300	233	N/A	120	267	N/A	-71	-54	278	216
Ireland	20	96	92	92	91	92	90	91	71	46
Italy	99	175	N/A	227	99	N/A	147	65	123	80
Luxembourg	0	33	67	67	-[6]	67	67	-[6]	5	-44
Netherlands	0	-6	-81	-86	-11	-91	-86	-11	125	-39
Portugal	138	188	200	200	200	180	202	202	317	164
Spain	-23	329	77	59	51	77	61	53	108	3
United Kingdom	73	47	120	28	-39	66	58	17	38	-22

* [1]The choice of rate elements required some judgement. Full details are available from the author. Generally, Peak/Off-Peak refers to the changes in the highest and lowest charges per minute for: local usage, calls under 25km (shorter distance), and calls over the longest distance (longer distance, usually greater than 150km). [2]For international calls, 'high' and 'low' refer to the range of percentage changes in charges from, e.g. Belgium to other countries. [3]Calls to Austria, Belgium, Denmark, Finland, France, the FRG, Greece, Ireland, Italy, Luxembourg, the Netherlands, Norway, Portugal, Spain, Sweden, Switzerland, UK, and US. [4]NA - Not Available. [5]France, local changes are Paris only. For other countries taxes are included as specified in the original data sources. For Denmark charges are for the Copenhagen Telephone Company only, and for the UK, for British Telecom only. [6]No difference in step for usage charge.

The pattern for the United Kingdom shows clear evidence of rate restructuring. There are relatively large increases in exchange line components, local usage charges and inland calls over shorter distances. In the United Kingdom there are decreases for longer distance inland calls as well as for international calls to some countries. The data for the United Kingdom show the 'classic' model of rate restructuring, one that no other country in Europe demonstrates to the same degree. As shown by Table 2, local cost per minute increases in constant 1980 prices for peak calls are countered by smaller changes for shorter distance inland calls, and a substantial decline in charges for longer distance inland calls. Prices for international calls to some countries also show a decline.

These data confirm that rate restructuring is underway in the EC. For the most part as shown in Table 2, increased charges in current prices translate into decreases in real terms. There are only a few cases where rate restructuring either between local and long-distance telephone services, or for calls over shorter and longer distances, has resulted in real price increases.

This empirical exercise suggests that tariff restructuring has yet to run its course in EC countries. There are still opportunities for discussion on the tariff structures within each country that will best contribute to economic and social policy objectives. Furthermore, in the face of rapid technical change and increasing competition in the EC, there is room for consideration of alternative pricing strategies and the extent to which they imply benefits and costs for different users. Although overall trends suggest that prices for local telephone services (i.e. installation, rental and usage) are increasing, this is not the case in all EC countries. Even in the face of increasing investment in new telecommunication plant, some PTOs are maintaining pricing policies that do not imply a redistribution of the costs of accessing and using the telecommunication network to those for whom a disproportionate amount of use is local service.

Table 2: Patterns of Tariff Restructuring Telephone Service, Percentage Changes 1980/87, Constant 1980 Prices

Country	Exchange Line		Inland Calls/Min.[1]						Internat. Dialled[2] Telephone Calls/Min. Range of Price Incr/Decr.[3] (From Country to 18 countries)	
	Installation charges	Rental charges	Peak			Off Peak				
			Loc	Short Distance	Long Distance	Loc	Short Distance	Long Distance	High	Low
Belgium	-36	-21	N/A[4]	-13	-13	N/A	-13	-13	3	-40
Denmark	-14	4	N/A	-3	-30	N/A	-9	-31	-33	-55
France[5]	-79	-50	-13	-13	-19	-38	-38	-42	-35	-65
Federal Republic of Germany	-74	-18	-18	-19	-18	-19	-19	43	-18	-59
Greece	16	-4	N/A	-36	6	N/A	-92	-87	9	-9
Ireland	-35	6	3	3	3	3	2	3	-8	-21
Italy	-9	26	N/A	50	-9	N/A	14	-24	3	-17
Luxembourg	-31	-7	16	16	-[6]	16	16	-[6]	-27	-63
Netherlands	-15	-20	-84	-88	-24	-92	-88	-24	92	-48
Portugal	-33	-18	-15	-15	-15	-20	-14	-14	18	-25
Spain	-62	113	-12	-21	-25	-12	-20	-24	3	-49
United Kingdom	15	-2	46	-15	-59	10	5	-22	-8	-48

* [1]The choice of rate elements required some judgement. Full details are available from the author. Generally, Peak/Off-Peak refers to the changes in the highest and lowest charges per minute for: local usage, calls under 25km (shorter distance), and calls over the longest distance (longer distance, usually greater than 150 km). [2]For international calls, 'high' and 'low' refer to the range of percentage changes in charges from, e.g. Belgium to other countries. [3]Calls to Austria, Belgium, Denmark, Finland, France, the FRG, Greece, Ireland, Italy, Luxembourg, the Netherlands, Norway, Portugal, Spain, Sweden, Switzerland, UK, and US. [4]NA - Not Available. [5]France, local changes are Paris only. For other countries taxes are included as specified in the original data sources. For Denmark charges are for the Copenhagen Telephone Company only, and for the UK, for British Telecom only. [6]No difference in step for usage charge.

Source: Calculated on the basis of Tarifica, Logica UK Ltd. 1980/87.

5. Conclusion

Two main approaches to pricing strategy appear to be unfolding. One seeks to initiate new advanced services primarily for sophisticated business users, leaving diffusion of new applications to the 'market'. The other attempts to encourage the development of service applications for the 'mass' end of the market by maintaining some aspects of traditional price structures (e.g. relatively low prices for access and local network use). How these divergent policies and pricing practices develop and their implications for the Single European Market could become an issue for more broadly based debate within the EC. There is a growing need to include a wider community of interest in pricing policy discussions. Voices, in addition to those of the largest business users, need to be heard. Tariff restructuring is necessary in the changing telecommunication environment. But restructuring should not be equated with 'inevitable' trends such as those which appear in the United Kingdom and some other EC countries. Apart from trends toward declining international tariffs for telephone service, experiences in each country are mixed. Different national telecommunication priorities continue to be incorporated in the pricing of exchange and inland services.

Reflecting technical change, national and international alterations in market structure and the increasing importance of telecommunication service applications in the economy, 'cost' - however defined - is beginning to play a more significant role in pricing decisions. However, this does not mean that prices for different service components will necessarily closely reflect their costs. There is likely to be continuing pressure to make the relationships between costs and prices more explicit in order to justify changing tariff structures both for economic and political reasons. Since the examination of cost and pricing policy is subject to national oversight, it seems unlikely that substantial convergence in switched service price structures throughout the EC is likely to emerge before, or for some time after, 1992.

References

BEUC (1988), *Consumers and the Telephone Services in the EEC*. BEUC/238/87, July, Brussels

Bonbright, J. C. (1962), *Principles of Public Utility Rates*, New York, Columbia University Press.

Clark, J. M. (1923), *The Economics of Overhead Costs*, Chicago, University of Chicago Press.

Commission of the European Communities (1987), *Towards a Dynamic European Economy: Green Paper on the Development of the Common Market for Telecommunication Services and Equipment*, (COM(87)290 final, June, Brussels

Commission of the European Communities (1988), *Comments on the Green Paper up to February 1988*, Brussels, XIII/41(88)

Commission of the European Communities (1988), *Implementing the Green Paper on the Development of the Market for Telecommunications Services and Equipment*, (COM(88) final, February.

Garnham, N. (1988), *Universal Service in European Telecommunications*, OECD, Paris, June.

Johanson, T., Lewing, D. and Rogerson, D. (1989), *Cost based Tariffs for Data Services: Positions and Trends in Europe*, London, Ovum Ltd.

Kahn, A. E. and Shew, W. B. (1987), "Current Issues in Telecommunications Regulation: Pricing", *Yale Journal of Regulation*, Vol. 4, No. 2, Spring.

Mansell, R. (1988), "Telecommunication Rate Restructuring: Issues, Patterns and Policy Problems for OECD Countries", *prepared for the OECD/ICCP/TISP*.

OECD (1989), *Telecommunications Network-based Services: Policy Implications*, ICCP Report No. 18, Paris

Trebing, H. M. (ed.), *Diversification, Deregulation and Increased Uncertainty in the Public Utilities Industries*, East Lansing, Michigan.

Ungerer, H. and Costello, N. P. (1988), *Telecommunications in Europe: Free Choice for the User in Europe's 1992 Market*, European Perspectives Series, Brussels.

3

The Optimum Access Charge under Competition

Tsuruhiko Nambu

Faculty of Economics
Gakushuin University
Tokyo

1. Introduction

The purpose of this paper is to analyse the socially optimum access charge when an incumbent and newcomers are operating in the telecommunications market under regulatory authority. In every country cross subsidization between local and long distance market used to be justified from the aspect of income redistribution. And now this kind of cross subsidization is one of the keenest issues among countries where competition is introduced into telecommunications market. Competition always emerges when there exists an attractive profit opportunity. Under the traditional regulatory regime the long distance market provided a surplus fund to keep local rates lower and hence this market is naturally a target for newcomers to exploit profits. As new entrants increase their market share, so the profit of the incumbent is reduced. This will make it difficult for the incumbent to continue subsidization to the local market. So called "rate-rebalancing" is an inevitable result.

Rebalancing rates brings about two different effects in income redistribution. In the local market economic welfare of telephone users gets worse because of price increase and in the long distance market the opposite is true. But it is not certain whether the total welfare of telephone users is increased or decreased through this rebalancing. It is easily anticipated that long distance telephone users get better off as the market share of newcomers increases. On the other hand local rates must be raised by the decrease of subsidy which makes local telephone users worse off. In short, new entry into the long distance market renders different effects upon two kind of users. Then comes the question whether there exists an optimal rate of entry which maximizes the total welfare of both users and, if it exists, if such optimal entry can be obtained automatically. This is our focus of analysis in this paper.

In our paper "regulated" competition in the Japanese market is analysed based on a simple theoretical model which can also be applied to the European situation. In order to enable a comparison between Japan and Europe an institutional framework of Japanese telecommunications industry is slightly sketched here.[1]

In 1985 NTT (Nippon Telegraph and Telephone Co.) was created through privatisation of a public monopoly and entry into telecommunications markets was also permitted upon approval of the Ministry of Post and Telecommunications (here after MPT). Three firms entered into the long distance market between Tokyo and Osaka. These newcomers are called NCC (New Common Carrier) and are given some advantages or protection under the name of "industrial policy" of the MPT. The most stringent measure for protection is price regulation. The price structure of long distance telephone is a two-tier system. NCC's tariff is 20 percent lower than NTT's through price regulation. The quality of services provided by NCCs is regarded as the same with that of NTT and so the price differential does not correspond to quality difference. Naturally customers are always tempted to shift from NTT to NCCs. Two of the NCCs are already in the black in 1989 and the last one will be so in the near future although they started their telephone business two years ago. Their market share in the Tokyo-Osaka market is about 7% in 1989 and predicted to be much higher next year. This success clearly depends on the two-tier price control by the MPT.

---

1 For a more detailed discussion, see Tsuruhiko Nambu, Kazuyuki Suzuki and Tetsushi Honda, "Deregulation in Japan" in Robert E. Crandall and Kenneth Flamm (eds.), *Changing the Rules: Technological change, International Competition and Regulation in Communications* (Brookings Institution, 1989)

Another feature of the Japanese system is found in tight regulation of the local telephone rate. So far rebalancing between long distance and local rates has not been introduced and the local rate has been kept unchanged. NCCs do not pay an access charge to NTT though they pay the local rate twice when their customers are connected to NCCs point of interface. NTT is regarded financially capable of continuing subsidization to local telephone at least for the moment. "No access charge" is another kind of industrial policy of the MPT.

In what follows, we will analyse policy alternatives on strict premises to avoid wishful thinking. It is evident that NTT cannot subsidize local market forever facing competition from NCCs. If NTT is obliged to keep the local rate unraised, it will be in the deficit sooner or later and fail as an organization.

That is certainly not our choice. In order to avoid irrational choice we must raise the local rate unless we forbid entry of NCCs. We must also avoid the too optimistic assumption that NTT can continue cross subsidization because the long distance market is growing enough rapidly to make subsidization possible. This has not been the case in the past several years and we can not be so optimistic as to rely upon the most comfortable assumption.

2. Model

2.1 Local Market

Local telephone calls are subsidized by long distance calls which generate the necessary profits transferable to local market. The situation is depicted in Figure 1 where we assume a linear demand curve $p_X = a - bX$, p_X denoting the local telephone rate per call and X denoting local telephone traffic and a, b are positive constants.

Figure 1: Demand in the Local Telephone Market

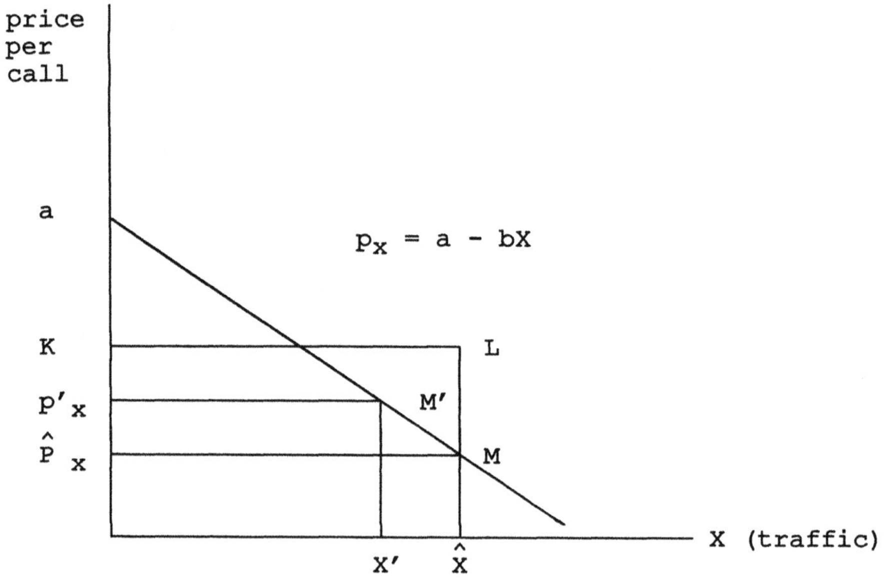

The regulated local rate is $\hat{p}_X$ and its corresponding traffic is $\hat{X}$. It is assumed that the supply of X creates the deficit of KLM$\hat{p}_X$ which is to be canceled out by the subsidy from the long distance market. The consumer's surplus is the traingle aM$\hat{p}_X$. When the subsidy is decreased, the local rate must be raised to, say, p'_X to cover the loss of this decrease. As local telephone demand is rather inelastic, there will be only a little decrease in traffic.

The local rate can eventually be raised until it becomes equal to the average cost of local telephone service. But judging from the present political situation in Japan, it seems to be as impossible as in the case of the United States. It must be noted that the average cost includes connection costs between the customer's premises equipment and the cable.

2.2 Long Distance Market

Under the MPT's regulation, NTT and NCCs provide services at different prices. NCC's prices are 20 percent lower their quality being equivalent to NTT's service. Accordingly, all the customers shift from NTT to NCC as long as NCC's supply is available. It is assumed that NCC's supply cannot exceed its capacity or the marginal cost of NCC becomes vertical at some level of traffic. NTT's market is the residual of long distance market after NCC's supply is deducted. This situation is shown in the next diagram, where again a linear demand curve $P_y = m - nY$ is assumed, p_y denoting long distance rate per call and Y denoting long distance telephone traffic, and m, n are positive constants.

Figure 2: Demand in the Long Distance Market

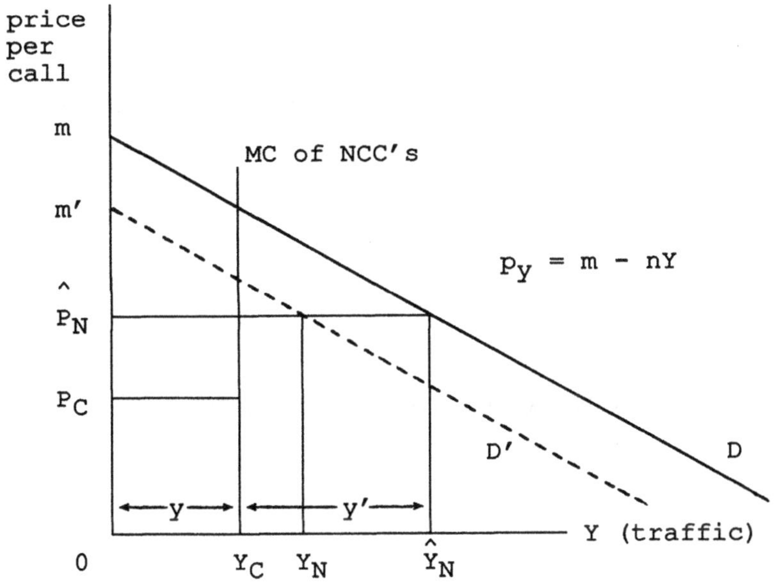

In the diagram, NCC's marginal cost is vertical at Y_C where their capacity is exhausted. NCC's supply is denoted y and the residual demand curve for NTT is shown by the dotted line m'D'. At price $\hat{p}_N$, which is higher than p_C, NTT can provide Y_N. Now the long distance market is divided into y and OY_N or $O\hat{Y}_N$ - y labelled y'. $O\hat{Y}_N$ is the total long distance market size at the regulated price $\hat{p}_N$.

2.3 The Behavior of NTT and NCCs

The behavior of NTT is just adaptive. It cannot control its market share because of price regulation and it can only provide traffic left for them. NCCs are maximizing their profits:

$$\pi_C = p_C Y_C - C_C,$$

where

π_C is NCC's profit function,

p_C is NCC's long distance price,

C_C is NCC's cost function, and

Y_C is NCC's supply.

NCC's cost is a function of Y_C and it can supply Y_C until it gets to the maximal capacity. The supply amount of NCCs is determined by maximizing profit π_C with respect to Y_C as in a competitive market, thus, being the solution of:

$$p_C = \frac{dC_C}{dY} = MC_C. \tag{1}$$

NTT's supply is OY_N or y' in the diagram. Accordingly, its profit π_N is defined as:

$$\pi_N = \hat{p}_N y' - C_N \tag{2}$$

where

π_N is NTT's profit function,

$\hat{p}_N$ is NTT's long distance price,

C_N is NTT's cost function, and

y' is NTT's supply.

2.4 The Social Surplus of the Long Distance Market

The consumer's surplus is the summation of NTT and NCCs users' surplus. This is shown in the Figure 3.

Figure 3: Social surplus of the long distance market

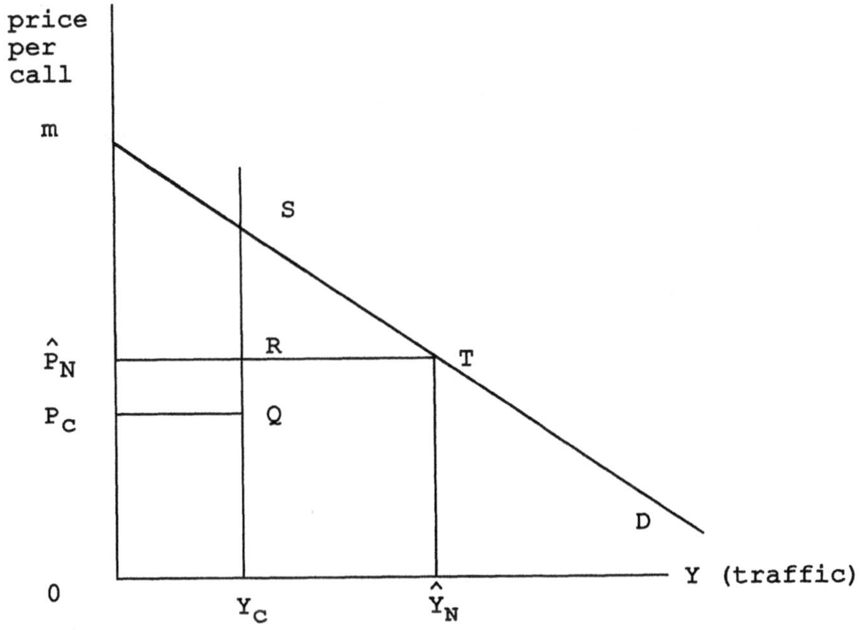

The consumer's surplus of NCC users is given by $mSQp_C$ whereas that of NTT users by STR. The total welfare is increased by $\hat{p}_N RQp_C$ compared to the old market where NTT provided $O\hat{Y}_N$ at the price $\hat{p}_N$. In addition to that, NCC's profit constitutes producer's surplus given by

$$\pi_C = p_C y - C_C \tag{3}$$

NTT's profit is given by (2) which is consumed for subsidizing the loss of local telephone market and is canceled out.

3. The Total Social Welfare

We are now able to calculate the social surplus by summing up the stakeholders surplus. The social surplus W is defined as

W = consumer's surplus in local market + consumer's surplus in long distance market + NCC's profit.

NTT's profit generated in long distance market is canceled out as a transfer for subsidizing local telephone users.

The consumer's surplus in the local market is equal to the triangle $aM\hat{p}_X$ minus the loss of NTT's profit in long distance market. It is additionally assumed that the welfare loss brought about by the reduced consumption is negligible because price elasticity of local telephone demand is almost nil. Now the consumer's surplus in the local market (CSL) can be written as

$$CSL = 1/2\,(a - \hat{p}_X)\,\hat{X} - (\hat{p}_N y - V_N\,(\hat{Y}_N) + V_N\,(y'))$$

where we define NTT's total cost C_N as

$$C_N = F_N + V_N$$

and

F_N is NTT's fixed cost,

V_N is NTT's variable cost.

In the long distance market, NTT has already sunk equipments and its fixed cost is constant whereas there exists a variable cost which depends upon the amount of traffic.

On the other hand the consumer's surplus in the long distance market is equal to the triangle $mT\hat{p}_N$ plus $\hat{p}_N RQP_C$. NCC's profit must be added to this consumer's surplus so that the total surplus in long distance market (TSL) is equal to

$$TSL = 1/2\,(m - \hat{p}_N)\,(y + y') + \hat{p}_N y - F_C - V_C.$$

NCC's total cost consists of fixed equipment cost F_C and variable cost V_C. They must construct new facilities to increase their market share which is in a sharp contrast to NTT that does not need new facilities. Accordingly, NCC's F_C must be treated variable when we compare the total welfare by choosing alternate traffic volume.

Finally we have the total surplus W as follows:

$$W = 1/2\ (a - \hat{p}_X)\hat{X} + 1/2\ (m - \hat{p}_N)\ (y + y') - \hat{p}_N y$$

$$+ V_N(y + y') - V_N(y') + \hat{p}_N y - V_C(y) - F_C(y) \qquad (4)$$

$$= 1/2\ (a - \hat{p}_X)\hat{X} + 1/2\ (m - \hat{p}_N)\ (y + y')$$

$$+ V_N(y + y') - F_C(y) - V_N(y') - V_C(y)$$

We must note that the first three terms are constant and F_C, V_N and V_C are variable when we change the rate of entry or y. The shape of W depends upon these factors and we can analyse this function by differentiating W with respect to y:

$$\frac{dW}{dy} = \frac{dV_N}{dy} - \frac{dF_C}{dy} - \frac{dV_C}{dy} \qquad (5)$$

Several possible shapes of the W function are depicted from Case 1 to Case 3 upon plausible assumptions below.

Figure 4: Total Social Welfare and the Rate of Entry (Case 1)

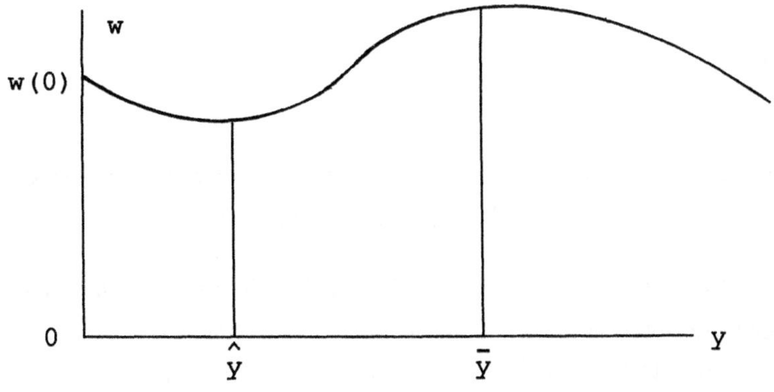

Figure 5: Total Social Welfare and the Rate of Entry (Case 2)

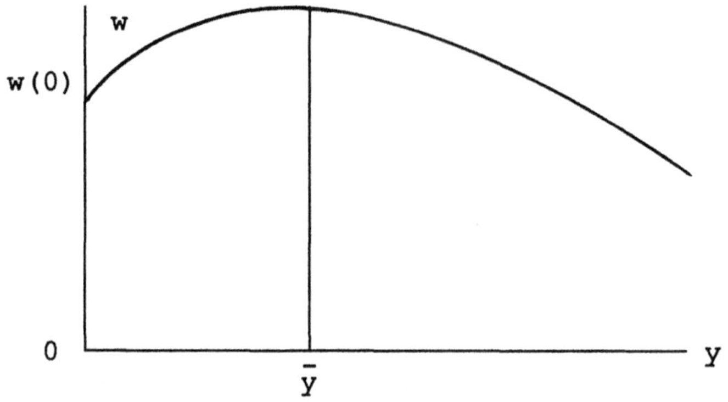

Figure 6: Total Social Welfare and the Rate of Entry (Case 3)

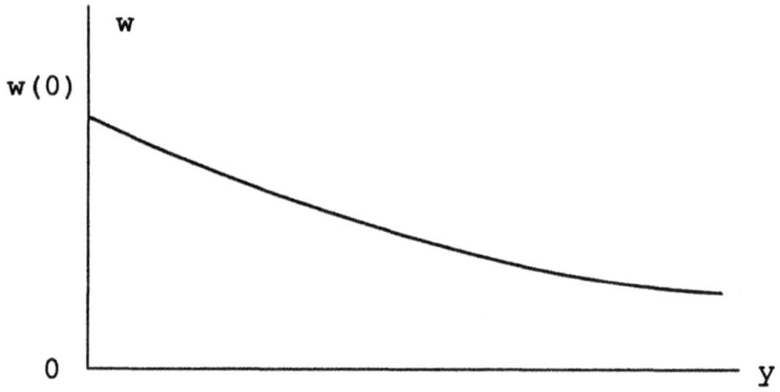

In case 1 the marginal cost of NTT is smaller than that of NCC up to $\hat{y}$ and after NCCs obtain $\hat{y}$, their marginal cost becomes lower than NTT's marginal cost.

This can happen when NCCs realize minimum efficient scale at $\hat{y}$. On the other hand NCCs must encounter increasing marginal cost when their market share exceeds $\bar{y}$. This is the case when the size of NCCs becomes equal to NTT (e.g. NCC's share reaches 50 percent) in the long distance market. NCCs can depend upon NTT's network before they get large enough. But if they acquire enough market share, it is compulsory that they should construct their own facilities preparing for peaks and for reserves. This would let NCC's investment rise sharply.

Case 2 and 3 are special cases of Case 1. In Case 2 NCC's marginal cost is lower than NTT's at the outset and it will exceed NTT's marginal cost when their share reaches the critical level of $\bar{y}$. On the other hand, Case 3 corresponds to the situation when NCC's marginal cost is always higher than NTT's.

The last case might well be disregarded because NCC's entry has no economic meaning. In case 1 and 2, we can find the maximal social welfare at $\bar{y}$, where the marginal costs of NTT and NCC are equal or

$$\frac{dV_N}{dy} = \frac{dV_C}{dy} + \frac{dF_C}{dy}$$

This optimal condition can be shown as below in the Figure 7, where NCC's supply is measured from left to right and NTT's supply from right to left.

Figure 7: The Optimum Condition for the Rate of Entry

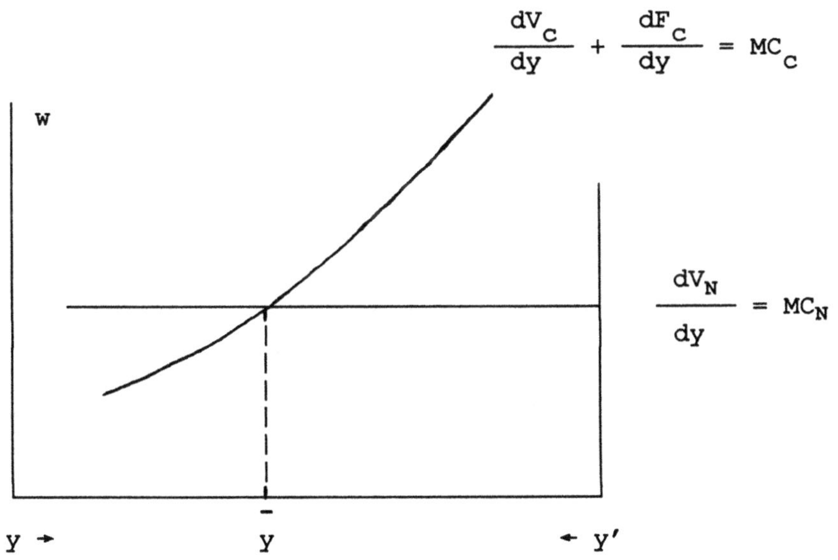

4. Access Charge: Two Viewpoints

According to the analysis in the previous section, we found that there usually exists an optimal rate of entry which maximizes the social surplus. We must now analyse the system which can bring us to this social optimum through regulated competition. It can be shown that the social optimum is not attained automatically when NCCs are free to enter into the long distance market. NCCs maximize their profit when their telephone rate p_C is equal to their marginal cost (see section 2.3)

$$p_C = \frac{dC_C}{dy} = \frac{dF_C}{dy} + \frac{dV_C}{dy} = MC_C \tag{1}$$

One obtains the optimal rate of entry from

$$\frac{dV_N}{dy} = \frac{dF_C}{dy} + \frac{dV_C}{dy} \tag{5'}$$

The system of (1) and (5') can be depicted in the diagram below.

Figure 8: Optimality conditions (1) and (5')

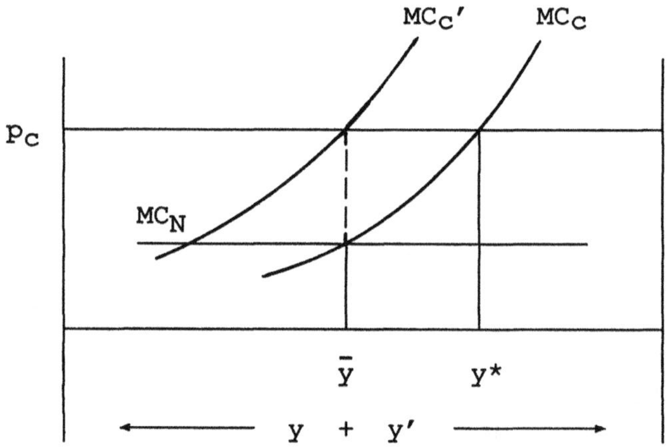

As is shown in Figure 8, NCC's price p_C is usually higher than NTT's marginal cost. We must recall that NTT's price p_N is about 20 percent higher than p_C thus making it possible for NTT to earn enough profits for cross subsidization. Although NCC's price is lower than p_N, it is very plausible that p_C is higher than NTT's marginal cost.

Now when p_C is higher than MC_N, the profit maximizing entry by NCCs occurs at y* where p_C is equal to MC_C. As far as MC_C is increasing, y* is larger than $\bar{y}$ which gives us the social optimum. In order to realize $\bar{y}$ we need some sort of government policy or taxation toward NCCs. The most popular remedy for this situation is lo levy a tax at the rate of t on the production of entrants. If we choose an appropriate rate of tax t*, the marginal cost of NCCs can be shifted to MC_C' which intersects p_C at $\bar{y}$. The optimal rate of tax t* is shown as

$$t^* = p_C - MC_C(\bar{y})$$

With the knowledge of MC_C and MC_N, we can calculate the level of $\bar{y}$ and this enables us to obtain t*.

Finally, we reached a system of attaining the social optimum through a usual method of taxation. We will call this optimal tax rate r* as "access charge". This concept of access charge is quite different form the traditional one adopted in the United States. The U.S. access charge is a kind of additional rate to cover the loss of local telephone provision by regional telephone companies. It is levied upon long distance users and carriers whereas local telephone users pay subcriber line charge (SLC).

The long distance and local users share the burden of deficit caused by insufficient increase of local rates. The SLC is an ad-hoc political compromise which accounts for the rise of local rates. On the other hand, in our optimal access charge system local rates must rise in accordance with the loss of NTT's market share. This increase of local rates is socially optimal in the sense that the rate of entry is determined by the access charge on NCCs.

We can find out the necessary amount of access charge by using a simple diagram as in Figure 9. For simplicity it is assumed that NTT's average cost is equal to the marginal cost MC_N which is horizontal. NCCs are taxed at the rate of t and the amount of tax is the area of $P_C\phi\kappa\gamma$ which is smaller than the loss of NTT, $P_N\varepsilon\kappa\gamma$. At $\bar{y}$, MC_C is equal to MC_N and the social surplus is maximized. It is important to notice that the necessary tax payment of NCCs is smaller than the acutal NTT's loss in the long distance market. NCC's profit is reduced by the amount of this access charge and is transfered to the government. The government can use this revenue like subsidizing local users. That is up to the government's choice but the social surplus is always maximized through our access charge.

It should be noted that NTT's loss of profit is represented by the area $P_N\alpha\beta\gamma$ if there does not exist an access charge upon NCCs and NCCs are free to enter. In this case social welfare is decreased by the amount of $\varepsilon\alpha\beta\gamma$ - $\phi\delta\beta\kappa$ or $\varepsilon\alpha\delta\phi$.

Figure 9: Necessary Amount of Access Charge

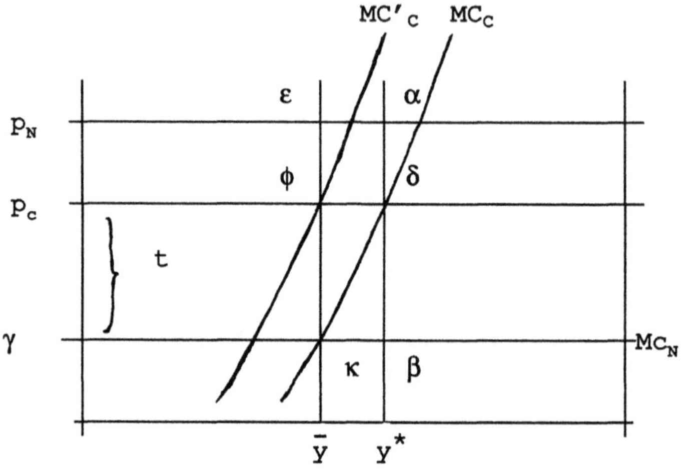

5. Concluding Remarks

In our discussion of policy alternatives we have avoided wishful thinking and assumed two things: rate rebalancing and no growth of demand in long distance market. It is quite easy to relax these assumptions and fly up to the possible competition where all the stakeholders get better off: no price increase in local telephony, a sharp decrease of long distance rates, financial success of newcomers and at the same time a powerful incumbent. This is often the scenario presented by reformers when they wish to introduce competition. But these kind of thinking has often nothing to do with the idea of economic efficiency or trade-offs in the institutional setting. From the economic point of view we should face the reality that some people (group) get better off at the cost of other people (group) when an established regulatory system is reformed because all the people have occupied the best position under the old institutional framework. Access charge is a typical case. From the U.S. experience we learn that all the stakeholders have their own say and find confusion about the appropriate distribution of fixed cost of local telephone service. Our approach in this paper is an attempt to find out unique solutions relying upon the rationale of economic efficiency or maximization of social surplus of all the telephone users. The analysis can be applied to other situations, for example where long distance market is growing more moderate rebalancing is possible through institutional arrangement or emergence of technological innovation which lowers the cost of local telephone and so forth.

References

Nambu, T., Suzuki, K., Honda, T. (1989), "Deregulation in Japan" in Robert E. Crandall and Kenneth Flamm (eds.), *Changing the Rules: Technological change, International Competition and Regulation in Communications*, Brookings Institution

4

The Regulation of Competition and Entry in Markets for U.K. Telecommunications Services

Martin Cave[1]

Faculty of Social Sciences
Brunel University
Uxbridge, Middlesex

Stephen Trotter[1]

School of Economic and
European Studies,
University of Hull
Hull

1. Introduction

This paper is concerned with the relationship between competition, entry and regulation in the U.K. telecommunications industry. Section 2 recounts developments in the regulation of entry and of the dominant firm since 1984; Section 3 discusses forthcoming regulatory decisions associated with the review of the 'duopoly policy', which will begin in November 1990, while Section 4 examines the relationship in theory and practice between the modified 'price-cap' form of regulation practised on BT and prospects for entry.

---

[1] We are grateful to N.J. Hartley, J. Peterson and D.J. Thompson for helpful comments. The views expressed are those of the authors alone.

2. Developments in Telecoms Regulation since 1984

The discussion focuses on three markets: basic service, mobile communications, and value-added services.

2.1 Basic Service

The initial basis of liberalisation of telecommunications in the U.K. is well known (see, for example, Vickers and Yarrow 1988). The 1981 British Telecommunications Act separated BT as a public corporation from the Post Office and introduced some competition in supply of customer premises equipment (except for the first telephone). It also gave the Secretary of State powers to licence other network operators, used in 1982 to licence Mercury plc to construct and operate a carefully specified service. The Act also allowed competition in the provision of value-added services, while excluding BT from supplying certain services. Finally, restrictions were placed on other providers' ability to resell BT capacity, principally on the ground that the low prices BT charged for private lines would introduce serious distortions if such resale were permitted.

Thus the beginnings of liberalisation pre-date the privatisation of BT undertaken in 1984. The twenty-five year licence awarded to BT at the time of the privatisation enshrined explicit prohibitions on anti-competitive practice and introduced the form of price control known as RPI-X. The anti-competitive provisions took the form of restrictions particularly on cross-subsidy of 'competitive' by monopolistic services, requiring separate accounting (with, of course, the attendant intricate cost allocation). These restrictions have been criticised as inadequate by, for example, Gist and Meadowcroft (1986). The institutional framework in which they are exercised by OFTEL and the Monopolies and Mergers Commission (MMC) also involves some jurisdictional complexities (Long 1988), but this aspect has not yet proved as serious a problem as it has in the United States, following the AT & T divestitive decision.

One of the major regulatory issues between BT and Mercury has been the extent of and charges for interconnection between Mercury and the BT network. Following disagreement between the two companies, the Director-General of OFTEL was obliged to use his powers to determine interconnection arrangements in 1985. Disagreements on charges have persisted, however, and arrangements for indirect connection to Mercury's services were not resolved until 1989. Agreements were also required covering domestic and international telex.

The price-cap formula, adopted in 1984, involved imposition of a minimum decline in real prices of 3 per cent per annum for five years on a basket of telecommunication services covering approximately 50 per cent of BT's output - access charges and local and trunk calls. In addition the Department of Trade and Industry secured an undertaking to restrict real increases in access and connection charges to 2 per cent per annum, and reviewed the increases in the relative costs of local as against trunk calls which took place between 1984-86. In concluding in 1986 that so-called rebalancing had not gone beyond what was justified by costs, the Director-General noted that he did not anticipate further real increases in local call charges (OFTEL 1986).

We return to examine the performance of BT and the relative success of Mercury below. First however it is useful to examine a number of recent changes in the regulations introduced in 1984. These relate to revision of the price control formula, changes in the rules governing resale and extension of the price-cap to private lines.

2.1.1 Revising the price-cap

OFTEL's discussion paper on revisions to the price-cap issued in early 1988, identified a range of issues, of which the key ones - given the form of price control would remain intact - were the duration of the control, its coverage, the size of X in the RPI-X formula, and incentives to maintain quality. (See Cave and Hartley 1989, from which this account is taken).

With the form of regulation now generally seen as a hybrid between price-capping and rate of return regulation, the duration of any price control formula obviously influences incentives to cut costs; constant revisions destroy incentives; on the other hand an immutable or long-lasting formula runs the risk of imposing unacceptable windfall gains or losses and lacks political or commercial credibility. This trade-off was resolved in favour of a four-year period.

The obvious criteria by which the coverage of the basket should be determined are the extent of competition and the ease with which financial results for different parts of the business can be separated. If two services cannot be separated then they should both be either included or excluded from the basket. In the consultative document OFTEL raised the issue of a substantial widening of the basket. The disadvantage of any widening of the basket is that less control can be exercised over individual prices - unless individual price-caps are also imposed. The advantage is that areas of remaining dominance outside the existing control can be brought into account. In fact, the basket was at that stage kept largely unchanged, although private lines were subsequently capped separately - see below.

Various approaches are available to the determination of the productivity growth target. Data on past productivity growth for the regulated company or levels of productivity or productivity growth for comparable companies are potentially useful. But data on performance of foreign PTOs are of limited value given the doubts about the basis of comparison, and conclusions are hard to draw from BT's own historical experience. For this reason most attention was paid to constructing a model of a hypothetical regulated company within BT, treating this as a separate entity.

OFTEL and BT were able to reach broad agreement on the form of a financial model for this hypothetical company, which was then used by OFTEL to produce projections. The key assumptions were prospective trends in macroeconomic variables (GDP, consumers' expenditure, the general level of prices); market demand (vital, because of economies of scale); BT's market share; the volume of factor inputs; and input prices relative to inflation; prospective investment levels; and, of course, the rate of embodied and disembodied technical progress.

OFTEL then studied the output of the model in terms of profit, capital employed, rate of return, growth of earnings, gearing and other financial ratios assuming different Xs. These results were assessed against broad expectations about a reasonable rate of return (given estimates of the cost of capital for BT as a whole) and expectations about the way in which new investment should be financed.

Finally OFTEL proposed, and BT agreed, that the new X for the period 1989-93 should be 4° (OFTEL 1988). It is interesting to note that if agreement has not been reached, the matter would be investigated by the Monopolies and Mergers Commission. Thus BT, like other regulated industries, effectively has a choice of regulators. This means that the balance of power between regulators and regulatee depends crucially on how the regulators interact.

The final issue to note is that of quality of service. The forms of quality incentives for a firm subject to various types of regulation are of major importance, and it is obvious that degradation of quality is a serious risk under price-cap regulation. The OFTEL consultative document raised the possibility of including an allowance for quality in the price-cap.

In practice OFTEL decided not to complicate the price-cap in this way - not least because of the substantial difficulties involved in identifying key indicators of quality of service, and then of setting explicit trade-offs between quality and price. OFTEL's approach has been to get BT to declare its own targets and then to amend its contract terms so that where the quality of the service delivered to one of its customers fails to meet declared targets, BT will pay a penalty.

2.1.2 The resale decision

Under the terms of the licences for BT and Mercury, restrictions on simple resale of private circuits (i.e. sales without a value-added element) would expire in June 1989, unless explicitly renewed. In January 1989 OFTEL issued a consultative document concluding with a recommendation of substantial abolition of the resale restrictions (OFTEL 1989). In July 1989 the Secretary of State announced the virtual abolition of the restrictions.

A principal reason given for initial restrictions on resale, against the advice of Beesley's 1981 report on ***Liberalisation of the Use of the British Telecommunication Network***, was the imbalance between price and cost for private lines. Prices were too low, and allowing resale would - in the absence of substantial price increases - enable resellers to undercut BT services, without exhibiting greater efficiency. Rather than allow the price increases, it was decided to prohibit simple resale.

In 1989 OFTEL concluded that the imbalance of prices had been reduced but not eliminated. This attenuated - or would attenuate by expiry of the price-cap in 1993 - the risks of inefficient entry. On balance, therefore, OFTEL concluded in its discussion paper, that 'a strong case does not now exist for delaying further liberalisation because of disruptive effects on the price structure'. It also pointed out that even if resale did threaten BT's capacity to cross-subsidise access or Mercury's to meet its license requirements, a transitional surcharge on resellers was an alternative 'price-guided' approach to dealing with the problem.

In the event, the Government accepted the case for virtual abandonment of the prohibition. As the necessary revisions to the licences have not yet been formally completed, it is not yet possible to see the effects of the liberalisation, and US parallels, where resellers have acquired a significant proportion of capacity for long distance calls are not necessary a good indication of U.K. developments.

The alternatives seem to be as follows. In one variant a company will lease private lines on certain routes and sell services directly to (normally quite small) customers. This approach may be profitable on particular routes where lumpiness in BT tariffs makes prices high relative to costs.

Alternatively, companies which already lease private lines for the purposes of their own communications may elect to sell spare capacity. Before the decisions were announced, one company (Plessey, which has however since been taken over), spoke of pursuing such a course. The scope for such resale is greater if companies combine forces to run a joint system. This already occurs in one case, where a single communications firm, EDS, runs a network for a number of companies. Such a combined system could in principle represent a new quasi-network, based on private lines and serving many large businesses. Any such developments, however, are likely some way to fall in the future.

2.1.3 Price-capping of BT's private circuits

The resale decision focuses attention upon prices of private circuits, where Mercury and BT compete in the supply of digital services, although only BT offers analogue services. As noted above, one major reason for prohibiting resale in 1984 was the excessively low price of private lines. By 1989, successive increases in BT's charges had brought prices up to a level which, according to OFTEL, was broadly congruent with costs. At the same time users of private circuits complained both about further prospective price rises and about quality of service.

In September 1989 the Director-General of OFTEL reported agreement with BT to impose a price-cap holding the real price of the company's private circuits constant for a four year period to 1993, concurrent with the RPI-4° regime for access and local and trunk calls. At the same time BT has agreed to compensate customers for delayed fault repair, and to set itself installation targets. The agreement allows BT to alter relative prices within the overall price-cap, but the Director-General will require the company to justify substantial price increases and will take subsequent action if he feels 'that BT is abusing its position of dominance' (OFTEL 1989a, p.11). The risk here is clearly that BT will cross-subsidise those services with which Mercury competes. The effect of the price-cap is to extend price control by slightly less than 10 per cent of BT's revenue - making a total of about 55 per cent.

On the issue of resale, the OFTEL report concluded that given the target of RPI-4° for telephone services provided by the PSTN, there was little risk even with the price cap of BT's public service losing business to an extent which would jeopardise the cross-subsidies necessary to maintain universal service and other social objectives.

It seems that the three regulatory changes outlined above - RPI-4°, resale and the price-capping of private circuits - are comparatively modest at least in their immediate effects, some in the direction of greater regulation, some in the direction of greater deregulation. The 'net' effect is hard to establish, though it is noteworthy that in setting or resetting the price-caps OFTEL has had to concern itself increasingly with BT's business plans and accounting procedures. This development is an inevitable consequence of the continuing lack of competition in the local network. We return to prospects here below. First, however, it is useful quickly to review the development of substantial levels of competition in two other telecoms services - mobile communications and value-added services.

2.2 Mobile Communications

The U.K. has already established a lead within Europe in the spread of cellular radio in an environment of keen competition, and now accounts for one-third of all subscribers in Europe. Moreover new services have been or are being licensed. The process has occurred in three steps.

2.2.1 Cellular phone networks

Two cellular phone networks were licensed to start operations in 1985, providing users with an opportunity to make calls from almost anywhere in the U.K., either directly to cellular phone users, or by interconnection with the fixed link networks. The two networks are operated by Cellnet (owned by British Telecom (60%) and Securicor (40%)) and by Vodafone owned by Racal. (This account is largely taken from Geroski, Thompson and Toker 1989).

The two network providers are however prevented by government regulation from providing cellular services directly to the customer, nor may they supply or produce communications equipment. Service provision is a competitive market entry into which is at the discretion of the network operators, who licence providers on the basis of financial resources and intended strategy. Owners of the two network providers can and do own service providers, but the latter must be separate subsidiaries. At first sight this aspect, combined with the licensing power of the network operators, appears to provide an opportunity for duopolistic control of the service market despite the requirement of vertical separation. In practice, the market share of the service providers owned by BT, Securicor and Racal is only 32%. The remaining market is shared by over forty other firms, with the pattern of entry and exit shown below.

	1985	1986	1987	1988
Entrants	23	22	14	8
Exit	-	4	8	6
Market participants	23	41	47	49

Source: Geroski, Thompson and Toker (1989).

The number of subscribers has approximately doubled in each of the three years between March 1986 and March 1989, reaching over 500,000 by the latter date - evenly split between the two networks. Before the announcement of potential competition from cheaper and more advanced systems (see below), the networks seemed set for an extremely profitable future. There have been problems with service quality, however, and OFTEL has thought it appropriate to improve consumer information by promising to publish a guide to service quality and monthly statistics on service congestion. More spectrum has also been allocated in the London area.

Geroski, Thompson and Toker have examined the impact of the vertical separation requirement on the structure of prices offered by service providers and by the network duopoly. They find a pattern in which service providers are encouraged by the connection incentives offered by network operators to sell equipment at low prices (at or below cost) and to set high usage charges in relation to cost. This produces a form of price discrimination between old and new subscribers; although a case can be made for this in terms of new network products, it is uncertain whether the actual outcome is welfare improving compared with cost-related pricing. Perhaps more importantly, vertical separation has been associated with (though not perhaps necessary for) competition in service provision.

2.2.2 Telepoint

Telepoint is a cordless phone system which enables subscribers to make outgoing calls if they are within 200 metres of a base station. It thus provides a less flexible but cheaper alternative to cellular telephones. Handsets are now available at £195; and user charges are be similar to those of call boxes (about 15p. per minute).

In January 1989 the Department of Trade and Industry licensed four telepoint operators which in contrast to the case of cellular telephones, will be allowed both to run a network and provide service. By the end of 1990 they will be required to adopt a common standard. This will allow any user to use any operator's base station. The licensees intend to start operation by the end of 1989 and one is already offering a commercial service. But progress has been slower than hoped, possibly because of subsequent development with PCN. Both Mercury and BT are members of successful telepoint consortia, though BT has been forced to assume a minority position and may not use its name or logo to market the service.

2.2.3 Personal communications networks (PCN)

The U.K. Government has recently sought applications to operate a new generation of flexible mobile phones using a higher part of the spectrum (1.7 -2.3 GHZ). These would offer a technology capable of providing both outgoing and incoming calls more convenient and smaller in scale than cellular phones. They will be micro-cellular systems, with cell sizes of about one kilometre. It is expected that development of the system will take several years, with a target starting date of 1993. Either two or three operators will be licensed, and it has already been announced that Cable and Wireless and its subsidiary Mercury would be one of them, subject to its making an acceptable proposal. BT and Racal were excluded from the competition. The operators chosen from the eight applicants will probably be required to fit in with standards adopted for DECT (Digital European Cordless Telephone). The question of mandatory vertical separation between network and service provision is not yet decided.

The implications for the public fixed links system of developments in radio communications have not yet become clear. Both cellular radio and PCN offer the possibility of by-passing fixed links and thus threatening BT's near monopoly of residential service. This will be facilitated as PCN operators are able to use radio to link the base stations within their networks, a right likely to be extended to Racal. The extent to which this happens clearly depends on relative costs, but the Director-General of OFTEL has noted that by the mid-1990s Britain would have four or possibly five companies providing full mobile networks: 'mobile systems will generate competition in local fixed-link communications where we have very little at present' (Financial Times, 11 July 1989). A less alarming prospect for BT and Mercury is that users would use mobile radio principally to access public telephone networks.

2.3 Value-Added Services

Value-added telecommunications services are those which offer facilities for altering and storing messages within the network which transmits them between senders and receivers. The basis of U.K. policy towards value-added services has been the promotion of competition as the means of providing the widest variety of services in the most efficient manner possible. Indeed, one of the first steps towards liberalising the U.K.'s telecommunications services market occurred when the then Industry Secretary, Sir Keith Joseph, relaxed the Post Office's monopoly over these services in July 1980. Since then, entry into the value-added services market has been relatively easy. Most operators of these services simply had to obtain a general licence - a Value-Added Data Services (VADS) licence - rather than individual licences on a service-specific basis. But this was accompanied by the prohibition on the sale of basic services to third parties.

There have been cases where the definition of basic services used in the licences is alleged to have created difficulties for some private network operators; in particular those offering Managed Data Network Services (MDNS). Although licence requirements were modified in 1987 to allow freer use of private circuits to provide data services, a division between basic and value-added services continued to be a feature of the licence.

Given that the aim of regulation was to allow competition in areas other than basic service, the "working definition of value-added services tended to be based on exclusion rather than inclusion; they were contained within a residual category which emerged as a result of only specifying the characteristics of basic services. Peterson (1989) has sought to develop a more satisfactory classification of services which focuses upon a product characteristics approach. This approach is necessary for demand analysis, but for regulatory purposes the alternative 'residual' definition was sufficient for a framework in which regulation encouraged competition as the means of promoting diversity and efficiency.

After 1980 many firms entered the market, mainly providing information services such as data banks and electronic mail. Several studies have shown the presence of competition in some markets (King 1986). One particularly important development is the growth of international trade in telecommunications services, now the subject of discussions at the Uruguay Round of GATT talks. Because the Uruguay Round talks have focused on value-added services, regulatory issues are likely to become more intertwined with trade policy issues.

The decision taken in June 1989 by the Secretary of State for Trade and Industry to permit resale of private circuits has meant that operators of private inland circuits can now offer basic services to third parties. As a result of this new-found freedom it is arguable whether the term "value-added" retains a role as a regulatory dividing line. The June 1989 changes seem to indicate that "value-added" is now simply a generic term for certain classes of services which exist within a wider range of competitive services that licensed operators are now in a position to offer under the Branch Systems General Licence (BSGL). It can be argued that the measures announced in June 1989 highlight a process under which U.K. policy is moving away from service liberalisation and towards infrastructure liberalisation; this is based on the view that services liberalisation emerges as a natural concomitant of greater freedom of use of circuits.

3. The Communications Infrastructure, Competition and the Duopoly Review

The previous section has outlined the development of competition in certain key areas of telecommunications services since 1981. While progress has been made in some areas, the dominant position of BT in provision of basic service remains fundamentally unaffected. Beesley and Laidlaw (1989, p.45) estimate Mercury's output in 1988 as £120 million, of which about one half is switched voice services. (The Cable and Wireless Annual Report for 1988/9 cites a revenue figure of £181m., but it is not clear how BT interconnection charges are treated; they are excluded in the estimate of £120 million). BT's revenues over the same period from the public telephone networks are estimated at about £9bn. Thus the impact made so far by Mercury is small, although it is now reported to be in profit; the Cable and Wireless Accounts do not however explain the basis of the calculation.

The policy of expanding competition adopted since 1981 has thus had limited success in reducing BT's market share, though its effects in making relative prices more efficient through 'rebalancing' are hard to gauge. The review of BT's and Mercury's duopoly in the provision of basic service to begin in November 1990 imposes the next crucial decision point. As far as the communications infrastructure is concerned, the key questions are: Should policy promote construction of a monopoly national grid, possibly regulated on a common carrier basis and licensed to deliver broadcast as well as telecommunications services? If on the other hand this policy is not pursued, how should competition be regulated in respect of the number of firms permitted and the range of services provided by each?

These issues are the subject of a report by the Communications Steering Group, a group of experts appointed by the DTI, entitled *The Infrastructure for Tomorrow* (though it should be noted that the views expressed are not those of the Government).

The report considered the case for two alternative policies. The first involves a clear Government commitment to support by subsidy or regulatory policy the construction of a national broadband fibre optic grid, covering most of the population but using other delivery technologies in remote areas. Such a grid would effectively displace competing networks and operate as a natural monopoly.

The Group noted, however, that even assuming high demand for cable television and high definition television, projections made its consultants suggested that during the period 1990-2010 a policy to install such a system would not be profitable in the absence of specific fiscal incentives. Moreover, most new services including HMTV could just as well with suitable compression be provided by the installation of an ISDN narrowband network.

The alternative approach outlined (and favoured) by the Group involves taking the best elements of a variety of delivery systems. 'This would entail a number of complementary networks co-existing and competing with each other to meet varying user demands with each able to provide the interconnection users require to its fellows'. The Group's illustration of this vision is given as Figure 1[2]. The regulatory implication is that attention should be focused on harmonisation of standards and the maintenance of interconnectibility and on preventing the abuse of the incumbent's dominant position.

More concretely the Group proposed use of what it termed 'one-way franchising' to promote local provision of one-way (non-interactive) services. This would involve offering renewable local franchises free to provide services to customers in an area through the delivery technologies they deem best - in the case of entertainment services, probably cable or MVDS. The franchisees might then move into the carriage of two-way services, including telephone service. In practice, however, recent Government decisions on the interrelation between cable franchises and MVDS have tended to work against this policy by allocating frequencies at the higher (40 gigahertz) level to MVDS. The choice of the higher frequency rather than the cheaper but less capacious 10-12 gigahertz range, for which the technology already exists, is likely to hamper the role of MVDS as a precursor and sustainer of cable.

2 The table "The Relationship of Networks" has been taken from the Communications Steering Group Report *The Infrastructure for Tomorrow* (ISBN 0 11 514663 6). Permission of reproduction of this table has been granted by the Controller of Her Majesty's Stationery Office and is gratefully acknowledged by the editors.

Figure 1

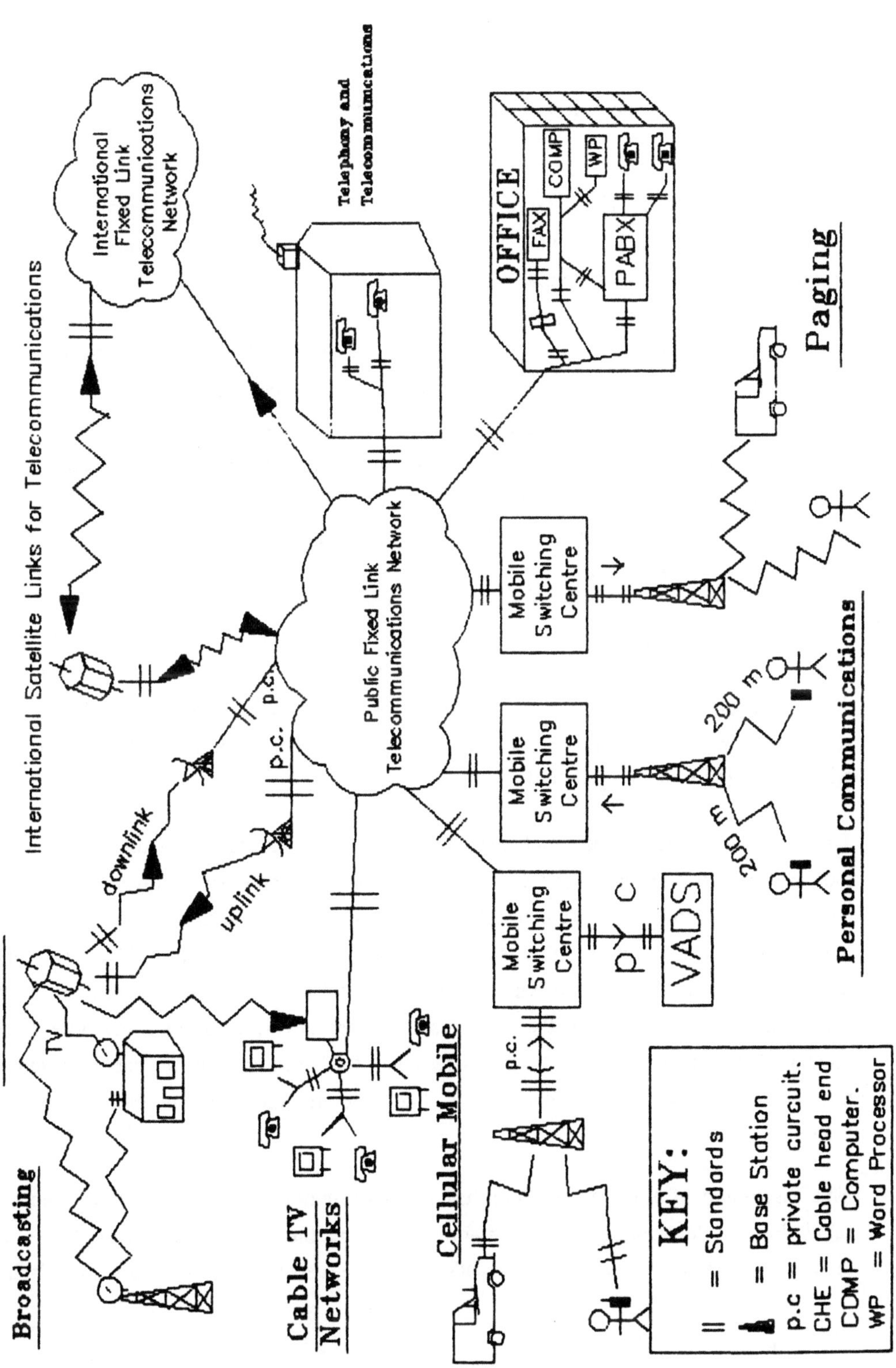
Satellite
Broadcasting
Cable TV Networks
Cellular Mobile
Personal Communications
Paging
International Satellite Links for Telecommunications
International Fixed Link Telecommunications Network
Public Fixed Link Telecommunications Network
Telephony and Telecommunications
OFFICE
FAX
COMP
WP
PABX
Mobile Switching Centre
Mobile Switching Centre
Mobile Switching Centre
VADS
downlink
uplink
TV
p.c.
p.c.
p.c.
p.c.
200 m
200 m
KEY:
= Standards
= Base Station
p.c = private curcuit.
CHE = Cable head end
COMP = Computer.
WP = Word Processor

The group recommendations are intended to be consistent with whatever emerges from the duopoly review. But the emphasis on the merits of competition and the aim abandoning asymmetric restrictions suggest a willingness to contemplate further entry into the provision of basic service. It is thus useful to consider in more concrete terms what form that might take, in addition to the mobile communications option noted in Section 1 above.

Beesley and Laidlaw (1989, pp.58-67) give a useful listing of the possibilities at national, regional and local level. At the national level they conclude that Mercury's experience will deter any potential entrant not already owning a network used for other purposes. British Gas is seen as the most likely candidate, though British Rail, which has recently set up a telecommunications division, is another, and further competition may come from the television transmission organisations, which the Government has proposed for privatisation. At the regional level there is the opportunity to develop a network rival to BT in areas not served by Mercury, using new digital technologies. The substantial investment would, however, be vulnerable to competitive changes in BT's tariff, though the authors note that 'BT's room for manoeuvre would be limited by its network configurations' (p.65).

The most interesting possibilities exist at the local level, where broadband cable systems have the option of providing local telephone service, and several of them have begun to do so. Franchises have now been issued covering per cent of homes in the U.K., although the number of homes passed is still small and the take-up rate still low at about 20 per cent (Cable Authority 1989).

There are, however, substantial regulatory problems facing the cable television industry's provision of telephone service, in addition to the commercial ones of securing an adequate customer base. BT is itself involved in about 30 per cent of systems, none of which offer telecommunications services. Secondly, only BT and Mercury can interconnect with cable systems which cannot link up directly to make a large joint local call area base, or deal directly with resellers. Moreover, local call businesses is comparatively unprofitable. For these reasons, the development of cable systems through telecommunications demand is likely to require considerable regulatory liberalisation.

This discussion of entry possibilities has focused attention upon the decisive role of regulation in determining the profitability and likelihood of entry, and Table 1 summarises the position in particular markets. The next section explores how much light can be shed on the link between regulation, entry and efficiency from the standpoint of regulatory economics.

But a second crucial issue which will determine the practicability of the strategy set out in *The Infrastructure of Tomorrow* is that of costs - the costs of offering alternative configurations of services and the costs of BT relative to those of potential competitors.

As noted above cross-country comparisons of absolute efficiency levels present well-known difficulties of measurement and interpretation, but such quantitative evidence as there is suggests that within a sample of European countries ranging in size from Denmark and Norway to Germany and Italy, BT's total factor productivity fall roughly in the middle (Foreman-Peck and Manning 1988).

Studies of the growth of total productivity yield divergent results. Beesley and Laidlaw (1989, p.39) estimate BT's growth of total factor productivity as 2.3 per cent per year over the period 1984-8, while Bishop and Kay (1988, p.45) have an estimate of 2.5 per cent of the same period.

Subissati (1989) however produces higher estimates in the range 3.1-4.4 per cent for 1984-8. This compares with a figure of 0.5 per cent per annum over the period 1978/9-85/6 (Molyneux and Thompson 1987) though the increase is likely to be due in part to short-term scale economies. While the higher estimates are broadly consistent with foreign experience, the lower figure of 2.3-2.5 per cent per annum is poor by international standards.

Finally, less formal comparisons of BT's prices (and, presumably, costs) with those of the other local telephone system in the U.K. (in Kingston-Upon-Hull) - and observations of comparative manning levels - suggest scope for productivity improvements within BT. The difficult judgement facing potential entrants however, is whether the emergence of actual competition would provide an incentive for BT strong enough to trigger cost reductions which would make entry unprofitable.

Table 1: Competition in Telecommunications Services: The Regulatory Framework

Service	Restrictions on Entry	No. of Competitors	Restrictions on BT	Restrictions on Mercury
Equipment sales	None	Many	Separate accounting required	Separate accounting required
Basic telephone service	Duopoly policy until November 1990	Two (plus local service from broadland cable) etc.	Price-cap regulation no undue discrimination etc.	None (except fair trading requirement)
Private lines	Duopoly until July 1989: resale soon to be permitted	Digital: BT and Mercury, + resellers Analogue:	Price-cap (from July 1989), no undue discrimination etc.	None (except fair trading requirement)
Mobile communications networks				
(i) cellular telephones	Restricted entry	Two	Co-owner of Cellnet	None
(ii) telepoint		Four	Co-holder of licence	Co-holder of licence
(iii) PCN		Two or three	Excluded	Holder of licence
Value-added service	None	Many	Accounting separating required	Accounting separation required
Broadcast signals	Control by licence or franchise of domestic suppliers	Four/five terrestrial channels plus satellite, cable and MVDS	Excluded, except by involvement in separate cable franchises	Excluded, except by involvement in separate cable franchises

4. Price-Capping and Entry

We now turn to the relationship between the system of regulation currently in place for telecommunications and the prospects for entry into the industry. The main part of this section discusses different possible levels of regulation, but first we look briefly at price-capping itself.

4.1 Price-Caps

Conventional rate of return regulation has been increasingly challenged in recent years by the method of price-capping, both in the academic literature and in actual practice, and the main characteristics of price-caps in a multi-product firm are fairly well-known by now. The mechanics are simple: a weighted price index for the firm's output is not permitted to increase by more than a given amount over the relevant period. In the BT case this is expressed in the well-known 'RPI-X' formula, by which the growth in the price of a basket of BT's services must be at least X per cent less than the rate of inflation. The basket of goods in the index may or may not include all the firm's products; for BT only about 55 per cent of its output is covered. (We should also note that the BT regulation is not a pure price-capping one. It is a hybrid rate of return/price-cap system, as one period's return on capital is an important input to the next period's value of X.)

Among the advantages claimed for the price-capping method of controlling monopoly utilities is the greater efficiency it promotes. Any cost reduction made by the firm is kept by it in increased profits, and there is no incentive to 'gold plate' by unnecessary capital expenditure to boost the rate base. A price-cap system is also much easier to administer, as the exact details of the firm's capital and the allocation of its costs between different services within the cap no longer matter to the regulator. Both disputes and regulatory commission manpower are therefore reduced. Last but certainly not least, in the inevitably political atmosphere surrounding the prices charged by basic public utilities, a price cap is appealingly simple and easily understood by the general public.

The best known theoretical consequence of price-capping is Vogelsang and Finsinger's (1979) result that a multi-product firm subject to a profit constraint will in certain circumstances choose a set of relative prices equivalent to Ramsey prices. Moreover a sequence of appropriate regulatory interventions will guide a firm starting from inefficient prices gradually towards Ramsey prices. The required action by the regulator is to set the gap at a level which removes any excess profit made in the previous period. Thus the firm reaps all the benefit of a change in a particular period, but the excess profit will be clawed back in future periods. The Vogelsang-Finsinger result depends on some fairly restrictive assumptions, but it does show the theoretical power of price-capping in certain cases. Other implications of price-capping - on product quality for instance - are summarised in Cave and Hartley (1989).

4.2 Regulation and Entry

The literature on price-capping and rate of return regulation has focused primarily on their effects on technical efficiency and relative prices. Less has been written on the link between different forms of regulation and entry, which is the main point here. About the nearest is Braeutigam and Panzar's paper on the issue of regulated versus non-regulated output (1988). They consider a firm which has a monopoly in the (rate of return) regulated sector and the option of also entering a competitive market. As might be expected the firm will seek to load costs onto the regulated output; it may also underproduce in the non-regulated sector (to keep up costs allocated to the regulated sector) and choose an inefficient technology. All these distortions are added to the general lack of incentive to minimise cost in the regulated sector. Although they are absent in a pure price-cap system, they operate with reduced force in a hybrid regime.

There is thus rather little previous work to discuss here, as the focus of the literature on regulation has not been on the issue of entry. Conversely the general work on entry has been within the traditional paradigm of unregulated profit maximisation. Our aim here is therefore to identify some connections between the two areas of analysis.

In principle there is a whole range of answers to the question of how to regulate entry. The first sub-section below discusses the two polar solutions of minimal and maximal regulation, which are predicated on particular assumptions about the technology and information sets available, and introduces an 'experimental' middle way of partial regulation. The latter is developed further in the next sub-section, and some more formal modelling is brought in in the final sub-section.

4.2.1 A Regulatory Spectrum

One possibility is to go for ***minimal regulation***, which can produce a satisfactory outcome if the technical conditions are right. To begin with let us abstract from any differences in technology between the existing firm and the one wishing to enter, and assume that both face the same costs.

The simplest case is that of the one-product firm. The first issue is whether the incumbent will engage in predatory pricing should entry occur. If this is forbidden - as seems likely - the next possibility is for the incumbent to practise limit pricing, setting prices below the short-run profit-maximising level in order to make entry unattractive to the potential entrant. If the industry is a natural monopoly the firm may earn positive profits by this route.

In practice it is more likely that the firm is a multi-product one, in which case the outcome here too depends on cost conditions. If there is a natural monopoly then under certain conditions there will be a set of sustainable prices equivalent to Ramsey prices and no entry will take place. Without natural monopoly entry will occur; and if the entry is inefficient then a game of attrition will follow until one firm drops out.

The conclusion from these relatively simple cases is that there is little need for regulation, except possibly in the predation case. Price-caps are unnecessary, and the threat of entry will lead to efficient, sustainable pricing. This analysis is applicable in some cases, but not if there are sunk costs. The presence of the latter means that the threat of entry is no longer as effective as before (see Vickers and Yarrow 1988, pp.61ff.), so that the chance of being able to manage with minimal regulation is much reduced. Sunk costs also affect how one firm expects the other to behave before and after entry.

We will see later that sunk costs are fairly pervasive in telecommunications networks, so that the minimalist 'hands-off' solution is not going to be generally applicable except at the level of service provision. Moreover we are not designing a system from scratch, starting from a blank sheet of paper and symmetric firms; the *status quo* in most European countries includes a very large and well-entrenched incumbent firm. The consequences of this are well illustrated by the Director-General of OFTEL: "BT has a strong advantage in some respects: the established size of its network and customer base gives it economies of scale in large measure and these cannot possibly accrue to MCL to anything like the same extent, in the near future. Consequently, competition will be possible only if MCL is given other compensatory advantages ..." (OFTEL 1989b, para. 1.21).

If competition would prove inefficient for technical reasons then one way out is to go to the other pole of a *central planning* solution. The optimum is calculated and then enforced, so as to mimic the outcome of an efficiently working competitive market. Bundling, two-part pricing or other tools could be used to achieve efficient prices, and there are no entry implications. The constraint on this, as Beesley and Laidlaw (1989, p.36) point out, is the traditional one on any form of central planning: "To mimic the results of competition involves a great deal of information and very sophisticated models". Even if the information problem could be overcome, however, a total planning solution is sufficiently out of tune with the political zeitgeist, both in the U.K. and elsewhere, for its adoption to be a political non-starter for the foreseeable future.

If central planning were possible, the regulator could leap straight to the efficient number of firms and moreover attain efficient prices by a suitable form of regulation. This may involve complex non-linear pricing schedules, other forms of price discrimination, and commodity bundling. But the risks of such a policy are that pricing rules which are efficient for a monopoly may also deter entry where it would be desirable. For this reason Beesley and Laidlaw argue that 'to judge BT's prices and to take relevant action, the prices must be 'unbundled'.... Only [then] can prices discipline the behaviour of both consumers and producers and the regulator obtain good information about costs' (p.36).

Between the two polar extremes identified above lies a third way: partial regulation. This is the system currently in use in the U.K., where each market is treated separately according to its particular characteristics. Three categories can be distinguished: regulated, protected entry and free entry. The first is those markets where competition will not produce a desirable outcome and regulation is needed. 'Free entry' covers cases, like those discussed above, where competition can be allowed a free rein. 'Protected entry' is an intermediate category for markets where competition may flourish if entry can be managed; it is comparable with the approach advocated in the report on *The Infrastructure for Tomorrow.*

The key to which category a particular market falls into is whether sunk costs are important. Telecommunications is a capital-intensive industry, and much of the capital used, such as the switching equipment in local and trunk exchanges, is highly specific (cable and ducting are arguably somewhat more adaptable). The industry is therefore characterised by a large element of sunk costs, so that entry into and exit from network services are far from costless activities: the market is not a contestable one in the sense of Baumol et al. (1982). The practical effect of this is that the entrant may fear being treated more harshly once it has been tempted into the market, due to the asymmetry between firms within and without the industry. When a potential entrant is deciding whether to become an actual entrant it has to forecast the future behaviour of the regulator: a tightening of the regulatory screws could mean that an entry that had appeared profitable is no longer so. In the U.K. there have been suggestions for some industries that a change of government could lead to changes in regulatory policy of this sort.

4.2.2 Partial Regulation in the U.K.

Thus the approach we favour is the one adopted in the U.K.: segmenting markets according to their particular characteristics and proceeding from there.

Some markets cannot support any meaningful competition, usually for technical reasons, and in these cases there is no alternative to regulation. In time it may be possible to relax this in some cases, but not always; water distribution is an example where it is hard ever to see any scope for efficient competition, given the infrastructure needed and likely technological change. In these cases regulation is likely to have to become tighter and more detailed over time, as the incumbent firms become smarter and more able to outmanoeuvre the regulator in the same way that OFTEL, despite its aim of promoting more competition, has been drawn more and more into regulating BT.

When the market can support it the regulator can stand back and let the various technologies compete, allowing free entry to anyone who wishes. The most likely cases of this are markets where sunk costs are small or non-existent so that the market is easily contestable. Examples here include the market for Value-Added Network Services (VANS) and provision of certain mobile communications services, given available networks. For Telepoint services, for example, there is a technical restraint that all users should be able to use all base stations, but sunk costs are relatively low and there is unlikely to be any natural monopoly.

The third category is intermediate between these cases. It covers markets where some guarantee has to be offered to induce new firms to invest the necessary sunk costs. The regulatory system works by saying "we will give you a guaranteed advantage over a certain period; if you don't take it up then there shouldn't have been entry in the first place". The most obvious example in the U.K. telecoms case is the trunk call market, where Mercury was guaranteed that the market would remain a duopoly for at least seven years after it entered (until at least the review beginning in November 1990). Whatever the level of Mercury's profits in that period no other firms will be able to enter. Other examples of such 'managed competition' include the licensing of two firms in the cellular telephone network market, four for Telepoint, and two or three for PCN, with a provision that the two existing cellular network providers can migrate to the PCN frequencies later.

Clearly one risk of the protected entry strategy is that any wrong decisions cannot be reversed without breaking the undertaking given to the firms involved and sacrificing the credibility of subsequent commitments. New technical developments, for example, might make such a revision desirable. The duration of the protection is thus critical. Too short a period deters entry, yet too long a period runs the risk of forgoing subsequent efficient entry and the development of collusion or regulatory capture.

Borrowing the terminology of statistics, the regulator can make either Type I or Type II errors. The former is to permit free entry where it should not have been allowed because the market is not able to support additional firms. This is likely to lead to a disorderly or even collapsing market. The Type II error is to protect a market when it should have been opened up to free(r) entry in the first place, or where subsequent technical progress means that further entry is desirable. In this case the cost of the wrong decision is the benefits that are foregone by giving the incumbent(s) an unnecessary advantage. The difficulty of course is to tell which markets are which, and any policy to reduce the chance of making one type of error will tend to increase the probability of making the other.

The chief weapons available to the regulator to promote protected entry are a mixture of (implicit) 'price' and 'quantity' controls. By imposing special licence requirements on dominant incumbents, the regulator can influence their costs and prices. By preventing incumbents from entering certain markets, the regulator can create new telecommunications firms. By offering only a limited number of licences for entrants, even in cases where further entry is technologically possible, the regulator can create potential rents. All these methods have been used by OFTEL and the DTI, and it is clear that attention is paid to the interrelations between markets. Thus in his *Annual Report* for 1988, the Director-General wrote (partly quoted above): ..."I do not believe that fairness (towards BT and Mercury) can be judged item by item. One has to take an overall view. BT has a strong advantage in some respects: the established size of its network and customer base gives it economies of scale in large measure and these cannot possibly accrue to MCL to anything like the same extent, in the near future. Consequently, competition will be possible only if MCL is given other compensatory advantages.... As the competitive balance changes over time, the balance of factors such as these will need to be kept under review and may need modifying in the interests of continuing evenness in competition".

The policy of protected entry thus differs from 'central planning' in its experimental approach and flexibility, even at the cost of blunting incentives. It differs, in the case of some markets, from minimal regulation by its insistence on looking at interrelations between services and the dominant position of the incumbent. The general approach has been to view markets as potentially open to entry rather than as natural monopolies from which entrants should be excluded, presumably on the ground that decisions to create monopolies are less easily reversed than attempts to promote competition: exit is easier than entry.

4.2.3 Towards a Formal Model

The ideal analytical strategy is to model the three-way game that takes place between the incumbent, the regulator and an entrant firm (one is enough), but this has not been done yet. In the meantime a first approach to analysing the relation between regulation and entry can be made in terms of a simple model, in which a firm produces two goods sold in independent (but non-competitive) markets. The situation is illustrated in Figure 2.

For concreteness the diagram takes these as local and trunk calls (subscripted L and T respectively). We assume that the (inverse) demand curves in the two markets are linear, taking the form:

$$P_i = a_i - b_i q_i \qquad i = L,T$$

and the cost function is:

$$C = F + (C_L + d_L q_L) + (C_T + d_T q_T).$$

Simple profit maximisation by a monopolist involves setting marginal revenue equal to marginal cost in each market separately, and produces outputs of:

$$q_i^* = (a_i - d_i)/2b_i \qquad i = L,T$$

and a profit of n^*; this solution is shown as A in Figure 2. The iso-profit curves are ellipses centred at (q_L^*, q_T^*). Under rate of return regulation - interpreting this as limiting total revenue to (1+r) times total costs - the constraint is also an ellipse, centred below and to the left of A and increasing in size with r. The firm may not operate within the ellipse: it will choose point B tangential to the highest available profit ellipse. A price-capping constraint amounts to a restriction that the chosen position must lie on or above a line like KL; now the firm chooses point C.

Other variations can be played on this theme. For example, a price-cap on say local calls only would be represented by a vertical constraint like MN, with a solution at point D.

(To prevent the diagram becoming too cluttered this has been constructed to be on the same iso-profit curve as C.) In this simple example there is no effect on the level of output in the other market.

This example shows that different forms of regulation generate different configurations of price and output by the regulated firm. When entry is possible the firm may respond by altering relative prices to discourage or deter it. Most obviously, this will take the form of cutting prices in markets threatened by entry and recouping lost profits in unthreatened markets. Other more subtle means, such as vertical restraints, will also be employed. The weapons at the regulator's disposal include the prohibition of cross-subsidy through the tightening of price controls and prohibitions on entry by the incumbent into new markets for related services in order to provide entrants with a secure base from which to attack the incumbent's monopoly.

Figure 2

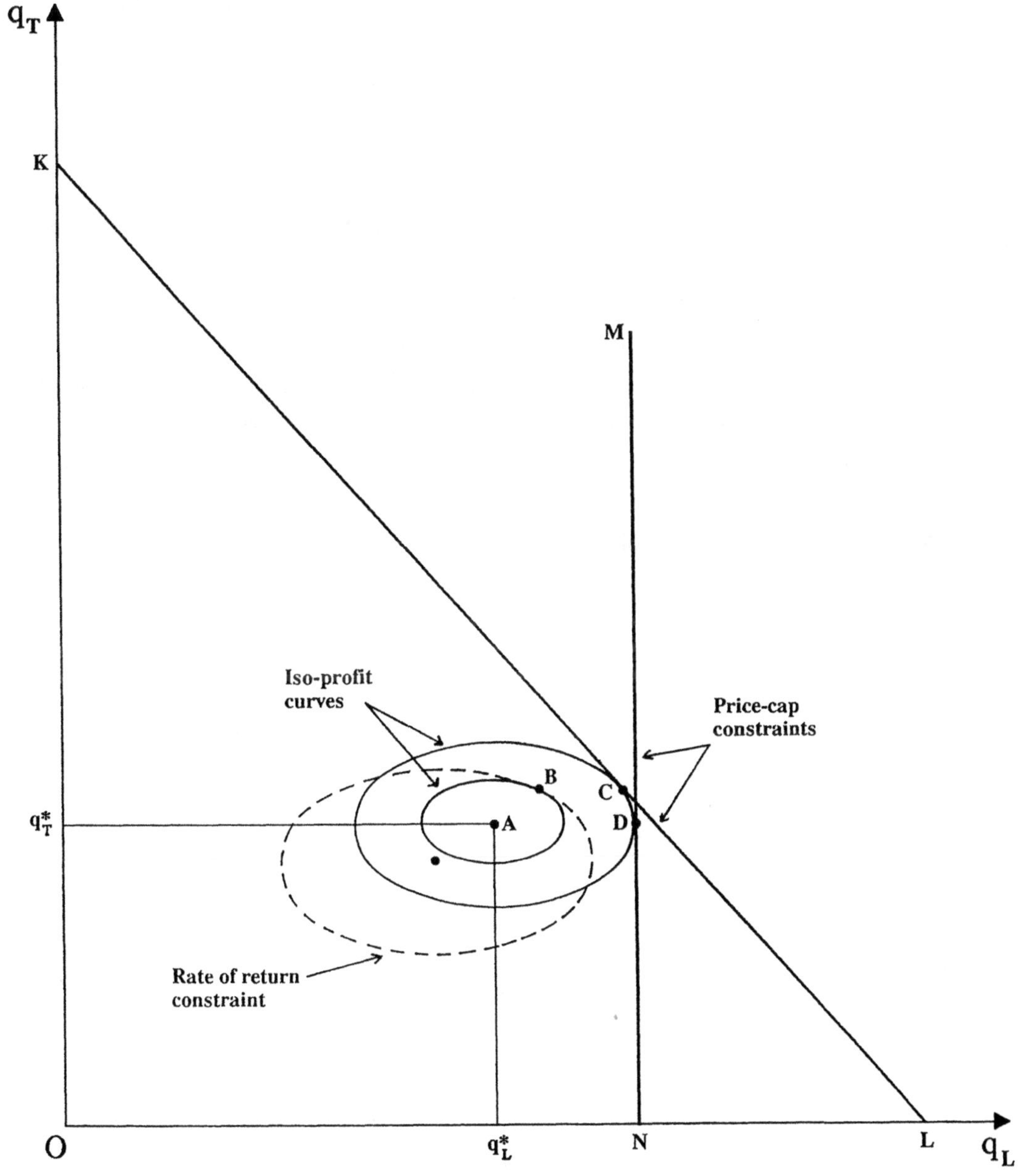
q_T
K
M
Iso-profit curves
Price-cap constraints
B
C
q_T^*
A
D
Rate of return constraint
O
q_L^*
N
L
q_L

5. Conclusions

This paper has been concerned with the relationship between regulation and entry in telecommunications, and in particular with the policies of what we have called 'partial regulation' and 'protected entry' in the U.K. In our view, the record of the policy so far is good. Competition has grown in several markets to an extent scarcely imaginable ten years ago, and unmatched elsewhere in Europe. Although some decisions are open to question - notably that concerning the number of licensees for Telepoint services - a determined but cautious policy of enabling competition can be discerned.

The crucial decisions, however, lie in the future at the duopoly review and associated decision whether to allow BT and Mercury to deliver broadcast services. Past policy has consisted of conjecturing the future efficient structure of the industry and then regulating with a bias in favour of more competition than less. In the past this policy has been applied largely to individual telecommunications markets or at least to the exclusion of broadcast services. This is no longer possible, and far more elaborate conjectures about the future industrial structures will be required, to apply the partial regulation approach at the duopoly review. We hope that the relevant regulatory bodies are already considering how to undertake this difficult task.

References

Baumol, W.J., Panzar, J.C. and Willig, R.D. (1982), *Contestable Markets and the Theory of Industrial Structure*, New York, Harcourt Brace Jovanovich.

Beesley, M. (1981), *Liberalisation of the Use of the British Telecommunications Network*, DTI.

Beesley, M. and Laidlaw, B. (1989), *The Future of Telecommunications*, IEA.

Bishop, M. and Kay, J. (1988), "Does Privatisation Work? Lessons from the U.K.", London Business School.

Braeutigam, R.R. and Panzar, J.C. (1988), Diversification Incentives Under "Price-Based" and "Cost-Based" Regulation, *Mimeo*, Northwestern University, December.

Cable Authority (1989), *Annual Report and Accounts 1988-9.*

Cave, M. and Hartley, N. (1989), "Regulatory Pricing in U.K. Telecommunications", *Conference on Telecommunications Costing in a Dynamic Environment*, San Diego.

Communications Steering Group Report (1988), *The Infrastructure for Tomorrow*, HMSO.

DTI (1989), *Phones on the Move: Personal Communications in the 1990s.*

Foreman-Peck, J. and Manning, D. (1988), "How Well is BT Performing?", *Fiscal Studies*, August, 54-67.

Geroski, P., Thompson, D.J. and Toker, S. (1989), "Vertical Separation and Price Discrimination: Cellular Phones in the U.K.", *Fiscal Studies*, November.

Gist, P. and Meadowcroft, S. (1986), "Regulating for Competition: The Newly Liberalised Market for Private Branch Exchanges", *Fiscal Studies*, August, 41-66.

King, N. (1986), "The Development of VANS in the U.K.", pp 449-463 of *Networks 26: Proceedings of the European Computer Communications Conference*, London.

Long, C. (1988), *Telecommunications Law and Practice*, Sweet and Maxwell.

Molyneux, R. and Thompson, D. (1987), "Nationalised Industry Performance: Still Third Rate?", *Fiscal Studies*, February, pp.49-81.

OFTEL (1986), *Review of British Telecom's Tariff Charges.*

OFTEL (1988), *The Control of British Telecom's Prices: a Statement.*

OFTEL (1989), *Further Deregulation for Business Users of Public Telecommunications Systems.*

OFTEL (1989a), *The Control of Quality and Prices of BT's Private Circuits.*

OFTEL (1989b), *Annual Report, 1988.*

Peterson, J. (1989), "Towards a Taxonomy of Telecommunication Services", CRICT, Brunel University, *Discussion Paper.*

Subisatti, E. (1989), "Some Evidence Regarding British Telecom's Performance Under Price-Cap Regulation", *Conference on Telecommunications Costing in a Dynamic Environment*, San Diego.

Telepoint (1989), "Lines of Communications", March, pp 34-7.

Vickers, J. and Yarrow, G. (1988), *Privatisation*, MIT Press.

Vogelsang, I. and Finsinger, J. (1979), "A regulatory adjustment process for optimal pricing by multiproduct monopoly firms", *Bell Journal of Economics*, Vol. 10, Number 1, Spring, 151-71.

Comment Part II

Godefroy Dang Nguyen

ENST Bretagne
Brest

The three papers that I have been asked to discuss, are very different in scope and background, although they all address a very important issue in telecommunications, that of pricing and regulation. To make it short, let's say that Dr. Mansell's paper gives an institutional analysis and description of tariffs evolution in telecommunications. Pr. Nambu achieves an interesting exercise in microeconomics, while computing optimal access charges of competitors to an incumbent's network monopoly. Finally, MM. Cave and Trotter give a very complete account of the competition policy pursued by the British Government in telecommunications so far, and present their own analysis of this policy, grounded on general economic principles.

It might be interesting to split the remarks I have to make to the three papers in several general headings:

1. Is economics in general, and microeconomics in particular, a useful instrument for the political decision maker to regulate parts or the whole of telecommunications?

   The answer of Pr. Nambu is clearly yes, while Dr. Mansell has several objections to the "esoteric" microeconomic world. MM. Cave and Trotter on the other hand, make an intelligent use of economics in their account of UK's competitive policy, which is neither purely formal nor deceptive.

2. Provided there are well established principles to justify the competitive policy in telecommunications, is there any contour to such a policy, or is it the simple outcome of constrasting and opposing self-interests, integrated by politicians' beliefs and career promotion?

   Here the discussion is between Dr. Mansell and Cave and Trotter, who seem to recognize that, although political decisions are difficult to take, the help of economists and consultants' work is of some value for the regulator, and may contribute to shape a first best policy.

3. What are the real motives for setting up tariffs, and how competitive forces may help to discipline the tariff-game?

4. To what extent, given the uncertainty surrounding an industry in quick evolution, can the regulator monitor the entry process? What is the degree of fairness in setting the number of authorized competitors, the level of access charges to the incumbent, the introduction of new products etc...

   Is it necessary to have a global vision of the future of telecommunications networks to regulate this activity or is it sufficient to apply general economic principles of competition?

We will discuss these items in the sequel.

The Usefulness of Microeconomic Models

Dr. Mansell has very strong reservations against microeconomic models to help the decision maker to set up tariffs. As she puts it: "pricing issues are debated in an esoteric theoretical world of optimality and second best solutions, which have little to do with the factors (social and economic) that decision makers actually consider in moving toward new structures". Conversely, Pr. Nambu exhibits one of these "esoteric theoretical" models, which the economists are so fond of. In my opinion, Dr. Mansell is perfectly right, but this will not prevent economists to design always more subtle and esoteric models! More seriously, such models are very useful in setting yardsticks, and general principles on which decisions should be evaluated. If we look for example to the UK experience in deregulation and the careful discussion around the RPI-X formula, one has to admit that OFTEL's decision to set up X to 4,5 % instead of 3 % has been taken with economic considerations in mind (see Cave and Trotter's discussion). In fact, BT and OFTEL have constructed a financial model of BT's activity, as the basis of their discussion and final compromise. Of course, it is possible to build up a microeconomic model to match any value of X, but the choice of a price-capping formula against a rate-of-return regulation was a political decision which was directly the consequence of an exercise in microecomomics (The Littlechild Report). One gets the overall impression, confirmed in Mansell's paper, that the more the telecommunications context is evolving towards liberalization, the more economic models are used in the public debate: In the US, the UK, and now Japan the outcomes of a political discussion on regulatory issues are less the result of a relative balance of power between stake-holders than a transparent compromise between the parties, who are trying to apply the lessons of general economic principles.

Once the usefulness of microeconomic models is admitted, at least as a mean to lay the ground for discussion, one has to revert to Pr. Nambu's tentative in the context of access charge. I have very much appreciated the clarity of Pr. Nambu's paper but I have two major reservations:

- The model is a static one,
- The model does not take account of reputational effects.

Let's take the latter point first. The experience of the US deregulation has shown that reputation is very important in this business. End users will be reluctant to switch from the incumbent, which generally has a strong reputation for reliability and quality of service (as this was the case for AT&T and, I presume, NTT as well) compared to a newcomer, even if the latter has lower prices. In fact, the only motive of the regulator to prevent the incumbent to lower his prices to the level of his competitors, is precisely because the regulator expects such reputational effects to act as barrier to entry, or at least to preempt the well functioning of competition. But is it possible to include reputation in Pr. Nambu's model? Yes, in a rather straightforward way.

We can suppose that demand to the newcomers is not

$$P_Y = m - nY$$

as in Pr. Nambu's model, but instead

$$P_Y = \mu m - nY.$$

That is, newcomers will not be able to bite into the whole demand of the incumbent, because, at any price, some customers will be reluctant to switch (Note that we suppose that this proportion of "irreducible" customers remains constant over the whole price range). Therefore, the newcomers will face a proportion of the potential customers of the incumbent.

The reputation effect does not change very much the calculations of Pr. Nambu for the optimal access charge, but it will obviously have an impact on the sustainability of competition as well as on the level of the optimal access charge.

On the other hand the reputation effect may shade over time, once the customers get used to the newcomer's ability to serve their needs. Hence, it might be interesting to analyze the dynamics of the diffusion of services by the newcomers. Pr. Nambu's model clearly does not address this issue.

Another point closely related to this, is the value of $(P_N - P_C)/P_N$, that is, the excess price imposed by the regulator to the incumbent. In the Japanese case, it is set up to 20% by the regulator to overcome reputational effects. If my intuition is correct, one should expect this ratio to vary over time. In other words, there is an optimal excess price ratio $(P_N - P_C)/P_N$, which should vary over time, in the same way as there is a static optimal access charge, as Pr. Nambu so brilliantly shows. In that case, it is naive to compute an optimal access charge while leaving the excess price ratio completely arbitrary. A full model, optimizing both values, is therefore necessary.

We can thus consider a fully dynamic model, where the regulator has two control variables: the excess price ratio of the incumbent with respect to his competitors $(P_N - P_C)/P_N$ and the optimal access charge, the latter in the lines suggested by Pr. Nambu.

One further hypothesis is needed, however: the dynamics of the reputational effect. One can assume that there is a learning effect which makes the parameter μ written above to increase over time: as long as customers become accustomed to the newcomers's ability to service them, the proportion of customers that the newcomers will reach is increasing. These dynamics will give the differential equation needed in the dynamic model. Although the dynamic model would not be fundamentally different from Pr. Nambu's static model, it is probably a step further in a realistic interpretation of the work of competitive forces.

Finally, there are some minor remarks: the model belongs to the category of partial equilibrium models, which is a restrictive, but not harmful hypothesis. Second, as Dr. Mansell suggests, the cost function of competitors may be particularly difficult to identify. In fact, one of the major problems she raises (how to allocate costs in a multiproduct context) is particularly hard to solve. Pr. Nambu's model holds only under the assumption of the regulator's perfect knowledge of the cost functions of the incumbent and the entrants, respectively. If this hypothesis is lifted, one has to resort to a principal/agent framework to devise an incentive scheme that might induce firms to reveal their true cost. Needless to say that such models, which are theoretically possible to set up, are more distant from reality because of mathematical tractability.

What is the Contour of a Pro Competitive Policy in the Telecommunications Context?

Only the paper of MM. Cave and Trotter address this issue, albeit in a narrow way, through the UK experience. But some hints are also given by Dr. Mansell, in presenting EEC Commission efforts to stimulate competition.

From the paper of MM. Cave and Trotter, we get the impression that some pro competitive measures work well (e.g. regulatory prohibition given to Cellnet and Vodafone to serve directly the customer, or corporate resale) and others do not (e.g. the authorization to sell local telephone services on cable TV premises) while some of them are workable political compromises (the RPI-X formula, or the discussion on the third carrier).

The lessons we can draw from MM. Cave and Trotter's account of the UK experience in deregulation are the following:

1. The competitive process is much easier to monitor and much more effective in areas where there is no asymmetric situation at the outset. In the case of mobile services for example, Cellnet and Vodafone nearly started from the same level and have achieved an outstanding level of penetration of services, due to competition. One can expect the same process to be repeated with Telepoint, where four competitors may be licensed, and with PCN, where two or three competitors will be established. The same applies in Value Added Service, a market which has boomed since the 1981 Act, without any restriction to competition. The final regulatory constraint, the interdiction to resell voice on leased lines, will be lifted in 1990, a good hint for the success of the British Government Policy in this area.

2. When the deregulation is aimed at cannibalizing well entrenched monopoly positions of an incumbent, the regulatory process works with many more difficulties. As reported by Cave and Trotter, Mercury, after 5 years of existence,

has taken only a minor part of BT's revenues (180 millions £ against 9 bn), while the potential market seemed particularly large. Also, to threaten local network monopoly with telephone services on cable networks is simply not a good idea.

The real threat to BT's quasi monopoly comes, according to the authors, from the possible take-off of the mobile services: Cellular paneuropean, Telepoint and PCN. They envisage that fixed telephone services may be substituted by mobile ones, provided by competitive consortia. However, their time horizon is the 21st Century. Also, the technical difficulties raised by universal mobile telephony are perhaps underestimated: what about the security of communication, the protection against fraud, etc...? Mobile systems seem more vulnerable to dishonest behaviors, than the fixed networks.

The intuition of our authors, grounded on some preliminary work of consultants to OFTEL, is that the future network in the UK will be a "geodesic network", to use an expression of Peter Huber, a consultant to the FCC: several networks, using different transmission facilities, and operated by different firms in a monopoly or competitive context, will be interconnected to build up the UK's communications cobweb. The centrally planned universal network of the 1960's will be replaced by a mesh of private and public networks, all of them interoperable. One may regret that Cave and Trotter do not address the fundamental issue of ONA/ONP, which seems to be absent in the British regulatory debate.

In the case of cannibalization (entry of newcomers in activities already supplied by the incumbent), the competitive process is clearly difficult to regulate. As suggested in Cave and Trotter, innovation is so rapid that the opening of competition in some market segments may be impossible to organize. This occurs in the mobile services, where competition is threatened by successive innovations: cellular, telepoint, PCN. Also, the pressure of technology blocks the standard setting process, and there is no way to prevent this: While the UK is pushing ahead a de facto standard, other European contries are muddling around a European standard. Moreover, the welfare costs of competition are not even mentioned by Cave and Trotter: Not only tariffs have been raised for local calls but also the quality of service has been downgraded in market segments where competition was not at work. OFTEL has been urged to design a regulatory mechanism to control the qualtiy of service of BT. This obviously has raised the cost of regulation. Finally, difficulties within BT (strikes) have generated welfare losses on users and business users in particular.

As pointed out by Cave and Trotter regulation is needed where there are sunk costs. In particular, when there is an incumbent the sunk costs of which are already bygone, some guarantee may be offered to entrants but, as suggested by my previous discussion of Pr. Nambu's paper, this guarantee is often difficult to set up. Also according to Cave and Trotter any wrong decision cannot be reversed without breaching into the regulator's own credibility. Finally, there is a strong incentive for government to support the newcomer once the competition is introduced. If the newcomer fails, because e.g. the incumbent is more efficient, the government may be wrongly accused to have devised a wrong policy, while the selecting mechanism of competition actually has nothing to do with politicians decisions.

The Issue of Tariffs

In the paper of Dr. Mansell, some figures are given which show that whatever the regulatory regime, tariff rebalancing is underway in most European countries. However, Dr. Mansell also suggests that there are two conflicting pricing strategies: one relates price to demand (Ramsey pricing) and the other accepts a certain level of averaging in order to recover costs.

The first pricing strategy is well suited to business customers who reject more and more averaging on basic service (telephone) but are ready to pay more for advanced services. The second pricing strategy is generally supported by Telecommunications Organizations (PTOs) which present it as a way to meet social goals (in the case of basic services) and to accelerate the diffusion of products (in the case of new services).

In a competitive environment the strategy of averaging is particularly vulnerable. In fact, PTOs are accused by their competitors to use averaging for cross-subsidizing services facing competition by monopoly services. However, the averaging strategy per se may be perfectly viable: most companies finance their growth in new activities with profits from "cash cows", that is activities where their position is close to monopoly. But the argument of fair competition is so strong in the European context that it may prevent PTO to implement averaging.

Another eventuality for a PTO is to sustain a lengthy and costly war of attrition in front of courts, as did AT & T in the 1970's, with the purpose to prove that either cross-subsidization does not exist or if it exists, it improves welfare. The implementation of the Green Paper in Europe may in the long run bring PTOs to such an end: once services are liberalized, contentions will be raised on cross-subsidization.

Dr. Mansell does not clearly take a position in the debate on 'value' against 'cost' pricing. Her main lesson is to suggest that there is a common trend in Europe towards rebalancing, but that there are still many variations depending on local conditions.

Conclusion

The three papers complement each other. They raise all the major issues of telecommunications regulation. They each have a distinct methodological background. They are valuable contributions to the open discussion.

Part III

International and Industrial Aspects of Telecommunications

5

Research and Development in Telecommunications: An International Comparison

Thomas Schnöring

Wissenschaftliches Institut
für Kommunikationsdienste,
Bad Honnef

1. Introduction

The telecommunications sector is characterized by massive structural changes due to changes in technology as well as in the regulatory environment. Research and development activities (R&D) become more and more important for the development of the industry. They determine the competitive position of firms and nations to a large extent. Both the equipment manufacturer and the network and service provider depend on their technological competence. In the past, the national R&D-systems of the industrialized countries in the telecommunications sector have developed separately. The systems that emerged were quite different from one country to the other. With the liberalization of the telecommunications markets and the increasing international competition (so far mainly in equipment sector) the different national R&D systems came under pressure and began to change. How do these national R&D systems look like? What are the advantages and disadvantages? How big is the R&D input of countries and firms and what is the output? What conclusions can be drawn from this for the future competitiveness in the international markets? These are questions the paper is about.

2. Some Theoretical and Methodological Remarks

2.1 Definitions, Demarcation and Data Sources

Although considerable progress has been made over the past years in the measurement of scientific and technological activities, there is no unique and superior measure of innovative activities and technological levels. Each has its strengths and weaknesses,[1] therefore we have used a range of statistical indicators (R&D expenditure, patents, scientific publications etc.) and complemented them with a series of interviews in each country. The judgements of the experts should help to bridge the inadequacy of the statistical indicators. The research project was a collaborative work of WIK and Fraunhofer Institute for Systems and Innovation Research (ISI), Karlsruhe.[2]

Definition of the subject "R&D in telecommunications" cannot be established unassailably. The demarcation used tries to deal consistently with diverse statistics and data sources and the definitions these contain. So what is telecommunications R&D? After perusal of various classification systems and the relevant literature, "Telecommunications" has been defined as a combination of 11 sub areas (compare Table 2.1). Other demarcations are conceivable.

Owing to lack of data, research statistics cannot differentiate between the individual fields within telecommunications. Consequently, R&D expenditure and R&D personnel can be quoted only for the field as a whole. Military-related R&D data are especially difficult to distinguish from civil data.

Delimitation of the individual sub-areas in telecommunications with the aid of patent symbols or classification symbols from the INSPEC bibliographical databank[3] used is a more accurate way of proceeding than using the verbal description in Table 2.1. Wherever possible, an entire sub-area defined by a classification symbol is allocated to one field only. In those cases, where this appears to be impermissible from the professional standpoint, apportionment is done by logical links with key words.

1 Grupp, H. (ed.) (1986), Problems of measuring technological change, Köln.

2 Grupp, H., Schnöring Th. (ed., forthcoming), Forschung und Entwicklung im Telekommunikationssektor - internationaler Vergleich mit zehn Ländern

3 INSPEC (Information Services for Physics and Engineering Communities) is a British bibliographical databank.

Table 2.1: Subfields of Telecommunications R&D

acronym	heading	contents
TH	information and communication theory, artificial intelligence	information theory, speech synthesis, general information and communication theory
WI	grid-based (wired) communication	grid-based transmission systems (incl. cables), telephone traffic (incl. telephone sets, private branch exchanges), switching, transmission technology electromagnetic converters (mouth piece), multiplexing for analog systems
RA	radio technique	antennae, aerial transmitting/receiving, radio communication systems (incl. satellites), noise reduction, control of oscillations, secret radio traffic
PI	general picture transmission	TV systems, image processing and recognition, cable systems for picture transmission, closed circuit TV
IM	image communication	facsimile (standing picture communication), moving picture communication (picture telephone)
DI	digital transmission	pulse technique, digital transmission (incl. telegraphy) analog to digital converter, pulse code modulation, encoding, digital transmission systems digital networks

FI	optical (fibre) communication	lasers for communications, optical cables, light control, light modulation, optological elements, optical analog to digital converters, optical semiconductor elements
HF	high frequency telemetry	wave guides, radionavigation, radar systems and equipment
ME	telemetering	traffic telemonitoring, telecontrol, telemetry
OE	other equipment (if not included elsewhere)	scanners, electrography, magnetography, data terminals, speech recognition apparatus, printers, plotters, electronic displays, holography, information storage
SO	services, software, policy	services, electronic mail, politics (laws, standards, norms) frequency administration, electronic media, TV systems applications, telecommunications in management, applications

Detailed analysis of R&D finances continues to be outstandingly problematical. At the heart of the matter is the so-called "Frascati-Definition" of R&D which states the following: "Research and experimental development (R&D) comprise creative work undertaken on a systematic basis in order to increase the stock of knowledge, including knowledge of man, culture and society and the use of this stock of knowledge to devise new applications".[4]

Activities to be excluded from R&D are teaching and training, scientific and technical information services and data collections, testing, standardisation and normalisation plus economic activities directed towards marketing (innovation) and production. The

4 OECD (ed.) (1981), The measurement of scientific and technical activities, "Frascati Manual", Paris, p.25

construction of prototype and pilot plants is to be included, design and building work only to the extent that they are required for the R&D phase and not the production process.

There is some argument as to whether software origination ("programming") equates to an R&D activity. On the one hand, the "state-of-the-art" in "software technology" is often improved, thereby elevating software generation to the status of a genuine R&D service. On the other hand, frequently only well established programming procedures are employed for writing and improving programs which cannot be rated as an R&D characteristic.

The most important supranational databank on R&D is managed by the OECD. From this databank, at the highest disaggregation stage, the R&D expenditure of economic sector ISIC 3832 can be obtained which include manufacture of radio and television sets and this is therefore too broad a concept for the basic definition here. So national sources of the most disparate kind need to be involved. Owing to gaps and inconsistencies in mostly heterogeneous data material and owing to nationally differing demarcations, examination and assessments of the material was unavoidable.

Overall, consistency trials show that the statistical data compiled are indeed reliable but display many gaps and inconsistencies.

Publications and patents can be interpreted as output indicators of R&D. ISI has analysed patent applications in the telecommunications sector at the US (USPTO) and at the German/European patent office (DPO/EPO) from 1975 to 1986 based on different databanks.

Apart from patent databanks which usually rely on official documents, bibliographical databanks are being set up in the private sector. The companies concerned have to prepare their own data files containing information on authors, subjects, addresses and key words. In term of consistency and completeness, therefore, the data held on bibliographical databanks are not of the same high level as comparable patent databanks. The choice of databank is thus critical to the investigation. Four internationally recognized bibliographical databanks were checked for their technical completeness. It was evident that the British databank INSPEC was clearly superior to other databanks. Therefore, INSPEC was chosen for biblographical analysis of publications in telecommunications from 1973 to 1987.

However carefully indicators are investigated and compiled, much important information about a national R&D system can not be gathered this way. A substantial part of the investigation was therefore an international comparative field investigation in which prominent R&D participants in the telecommunications field were questionned personally. The survey was conducted in 1988/89.

In the paper we present some results for the US, Japan, the Federal Republic of Germany, France and the UK. These five countries represent roughly 55 % of the world telecommunications market. Originally, the study was carried out for five countries more.[5]

2.2 The Model

In our analysis, a national R&D-system consists of four sectors: the network and service providers (carriers), the equipment manufacturers, the government sector, and the universities and other non-profit research institutions. These sectors carry out and/or finance R&D. The government sector mainly finances R&D for industrial and defence policy reasons, while the research institutions do research financed by the government and other sectors.We analyse the division of labour between these groups of actors and the dynamics of the system.

We assume that there exist distinct national R&D-systems which means for example that the intra-national flow of information and other types of intra-national interaction in the R&D-process is significantly closer than on the inter-national level. This assumption was certainly correct in the old world of telecommunications. It is still plausible because of the distinct history of each national network and the network character of telecommunications. But it is beginning to erode by the internationalization of manufacturers on the one hand and the liberalization of the equipment market on the other. International R&D collaboration among telecommunications manufactures is increasing rapidly, especially in Europe, and it is one objective of the telecommunications policy of the European Commission. The future will show to what extent collaborative R&D among competing firms and nations can work in the telecommunications sector.

For an economic analysis of the R&D system in the telecommunications sector the relation of carriers and manufacturers - vertical integration versus separated firms - is of special interest. Different arguments are brought forward about how this relation influences the amount, the orientation and the efficiency of R&D. First, a vertically

5 Italy, Sweden, Spain, Netherlands, South Korea

integrated system can better appropriate the results of R&D activities and will therefore invest more in R&D. The argument refers especially to research activities where one could not predict whether the service or the manufacturing side will be the principal beneficiary of the R&D activity. Second, in-house R&D is more efficient because of the superiority of intra-organisational methods of communication and other economies of scope between R&D for networks and services, and for manufacturing.[6] Both arguments have some merit and favour vertical integration. In United States versus AT&T both arguments were used by Bell in defence of the vertical integrated Bell System.[7]

But there is also some danger in vertical intergration. The R&D laboratories may acquire a sole orientation towards the needs of the carrier's division of the company which may be disadvantageous for the export business of the manufacturing division because their other customers may have different needs. This is especially a problem when the demand of the carrier's division is relatively small compared to the R&D overheads attendant to the development of new products.[8] On the other hand, it may be a disadvantage for the carrier's division to be solely bound to the company's own R&D and manufacturing division. Upstream R&D rivalry may prove to be a better choice from the carrier's point of view because it reduces the risk to pick an inadequate equipment. This argument gains influence in times of increasing technological change.

The arguments of economic theory remain contradictory and it becomes an empirical question which industry structure is best for R&D. Empirical evidence of the past seems to strengthen the first two arguments and the case for vertical integration. The US Bell System and the Canadian Bell System (Bell Canada and Northern Telecom) have been vertically integrated and in many European countries as well as in Japan we observe special relations and/or joint R&D activities between network providers and national manufacturers. But for different reasons these preferential relations have loosened in some countries. The US divestiture has separated the market of the BOCs from AT&T. Some national carriers in other countries started to introduce competing systems in their networks. National carriers follow more and more a policy of second sourcing. Increasing R&D costs force carriers and manufacturers - especially in Europe - to go

6 See for example, Olley, R. E.(1982), The process of Technology Transfer and Application in Telecommunications: A case study, in: Sahal, D. (ed.), The transfer and utilization of technical knowledge, Lexington, p. 175 - 203

7 Mowery, D.C. (1987), Forecasts of the impact of divestiture on Bell Telephone Laboratiories: An Assessment, paper prepared for the Telecommunications Policy Research Conference, Airlie House, Sept. 27-30, 1987, p.5; Noll, R.G., Owen, B.M. (1987), United States v. AT&T, The economic issues, Stanford University, California, Studies in Industry Economics, Discussion Paper No. 140, p. 58

8 This problem is mentioned by Olley in the case of Bell Canada and Northern Telecom, Olley, R.E.(1982), p. 187

international and by this loosen their traditional preferential relation.

Looking at countries with separated carriers and manufacturers one can distinguish two basic models when it comes to R&D for complex equipment systems like central office switches. In the first model, the national carrier is the leading actor. It does a lot of R&D itself, it is engaged in joint R&D projects with its traditional national suppliers and it finances most of the manufacturers' R&D activities directly by development contracts. The actual procurement of the equipment is done separately from the development. A limited number of national manufacturers which usually have been involved in the development of the system compete for contracts. This model has advantages and disadvantages. First, it is similar to the model of vertical integration with respect to economies of scale and scope in the R&D process and it solves the appropriability problem but it (usually) prohibits R&D competition on the national level. Second, it gives the national carrier's interests a high priority in the development of equipment systems which may be a problem for the export activities of the manufacturers. On the other hand the manufacturers have the advantage to compete internationally on the basis of production costs because most of the development expenditure is financed by the national carrier.[9]

In the second model the carrier has a relative weak position. It has only small in-house R&D resources. It is not involved in joint R&D activities and it does not finance R&D in the manufacturing sector directly but indirectly through the procurement of equipment. It gives the manufacturers an incentive to do specific R&D for its needs by restricting market participation to the winners of a first phase of a procurement procedure for a number of years. In this model, it depends on the price competition during the first years and the threat of potential competition thereafter to what extent the carrier pays for.[10] In this model economies of scale and scope cannot be realized as good as in the first model but there is R&D competition among the manufacturers. Manufacturers' interests and the market needs they face nationally and internationally have a higher priority. This can be an advantage in international markets. But they may have a disadvantage in international price competition. This depends on the price competition in their home market.

---

9 The manufactures have to carry R&D costs for the adoption of the national equipment to the foreign network environment.

10 Due to the technological competition of the manufacturers it is possible that the carrier has to pay for the development of more than one system. Assume a closed economy and two independently developping firms. In such a case the carrier would have to pay for the development of two systems if their is no cross-subsidization involved and both firms survive. Economically speaking, this is the price the carrier has to pay for R&D rivalry in his home market which reduces his risk to pick the wrong system.

The above two models are of course idealizations and we do not observe them undiluted in reality. In particular, the national R&D systems in Europe and Japan are mixtures of them.

3. International Comparison of R&D Expenditure

Information concerning R&D expenditure have been collected from national sources of the most disparate kind. Owing to gaps and inconsistencies estimations were unavoidable. Therefore, the results should be interpreted as "soft" data and not as "hard" facts. The information concerning the structure of each national R&D budget is more reliable than the information about the absolute national R&D budgets in telecommunications.

Table 3.1 shows the estimated national R&D expenditures for telecommunications.[11] It is differentiated between expenditures in the military and in the civil sectors. R&D in the military sector is R&D for communications financed by the defence department and R&D for communications in military sector financed by the manufacturers. The military sector has to be included for two reasons. First, R&D for military communication purposes can have spill-over effects for the development of equipment for civil use. Second, defence departments may fund R&D for communications equipment which can be directly used in the civil sector. The interviews of the Fraunhofer Institute provide examples of both effects, but it remained unclear how important these spill-over effects are. R&D in the military sector is large in the USA, France and the UK.[12]

The USA has by far the largest R&D budget for telecommunications, but more than one third is financed by the defence department (Table 3.1). Overall, the proportions of the national R&D budgets are in line with the proportions of the national GNPs.[13] In the civil sector, Japan's R&D expenditure is relatively large compared to that of the USA. This may be an advantage for Japanese firms in the future.[14]

---

11 We have got the impression that the US and the German R&D budgets are overestimated while that of Japan is underestimated, but we have got no reliable information to what extent. The US Figure is based on R&D in SIC 366.

12 Government military R&D expenditure as share of the whole national R&D budget in 1986: USA 38%, UK 24%, France 20%, Germany 6% and Japan 2%. Source: Fraunhofer-Institut für Systemtechnik und Innovationsforschung, Karlsruhe

13 The proportions of GNPs in 1986 are USA 100, Japan 40, Germany 21, France 17 and UK 16.

14 This is also the case for Germany. But we have the impression that R&D in the military sector is under-estimated in Germany and that some of it is included in the civil sector.

Table 3.1: National R&D expenditure for telecommuncations 1987a)

	USA	Japan	Germany	France	UK
1. R&D expenditure for telecommunications (billion US $)	13	4.7	2.5	2.1	2.1
a) in the military sector (billion US $)	5	0	0.2	1.0	0.7
b) in the civil sector (billion US $)	8	4.7	2.3	1.1	1.4
2. R&D expenditure for telecommunications in % of national R&D	10	10	10	12	13

a) Estimates based on national R&D statistics, government budgets and other sources. National currencies converted into US $ by current purchasing power parities used by the OECD for the field of science.

Figure 3.1: R&D for civil telecommunications 1987
Share of carriers' R&D expenditure

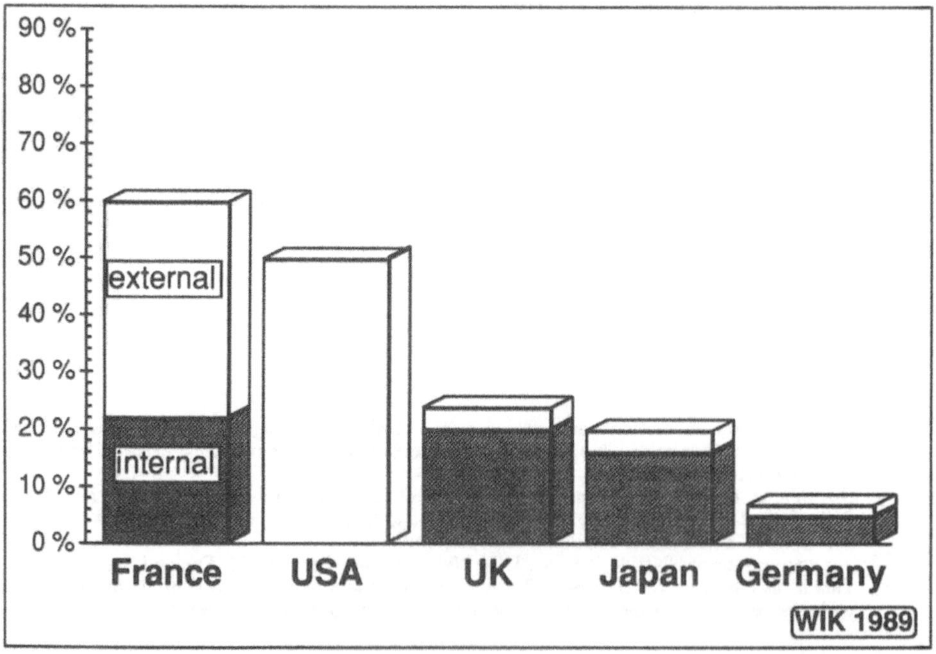

All countries spend an astonishing large share of their national R&D budget on telecommunications. It varies from 10 % to 13 %.

In the following we will restrict our analysis to the civil telecommunications R&D budget. Figure 3.1 shows the shares of carriers' internal and external R&D expenditure in the five countries.

In the USA, carriers perform half of the national R&D. This is not surprising because the vertically integrated carriers ATT and GTE are counted as carriers. Therefore, the US situation cannot be directly compared to the other countries where carriers are not vertically integrated.[15]

The French carrier France Telecom finances an even higher share of the national R&D than the US carriers. It finances almost half of the development activities of the manufacturers by development contracts. Internal and external R&D expenditure give France Telecom the leading role in the French R&D system.

In the UK and Japan carriers finance between 20 and 24 % of the national R&D. Most of the expenditure is for internal R&D.

Germany is the extreme case. Internal and external R&D expenditure of the Deutsche Bundespost (DBP) account for less than 10 % of the national R&D.

All governments subsidize civil telecommunications R&D to some extent (Figure 3.2). Money goes to universities, research institutes and manufacturers. The share of civil government money is relative small in all countries. It is interesting to notice that the often stated leading role of the Japanese government (MITI and MPT) in the national R&D system is not based on financial flows but other mechanisms.

Manufacturers finance most of the national R&D expenditure in Germany, Japan and UK (Figure 3.3).[16] The manufacturers' share is substantially lower in France, where manufacturers finance less than 40 % of the national R&D. It is likely that these differences have effects on international competitiveness in the world market.

---

15 For this reason, we have choosen a different colour for the USA in the figures.

16 For Japan and Germany, this is in line with aggregated national R&D numbers. In the case of the UK the result is surprising. Especially because the dominant manufacturers GEC and Plessey report that the share of their own funded R&D is just 40 % and 30 % of their whole R&D budget. But these numbers refer to all business areas.

Figure 3.2: R&D for civil telecommunications 1987
Share of government R&D funding

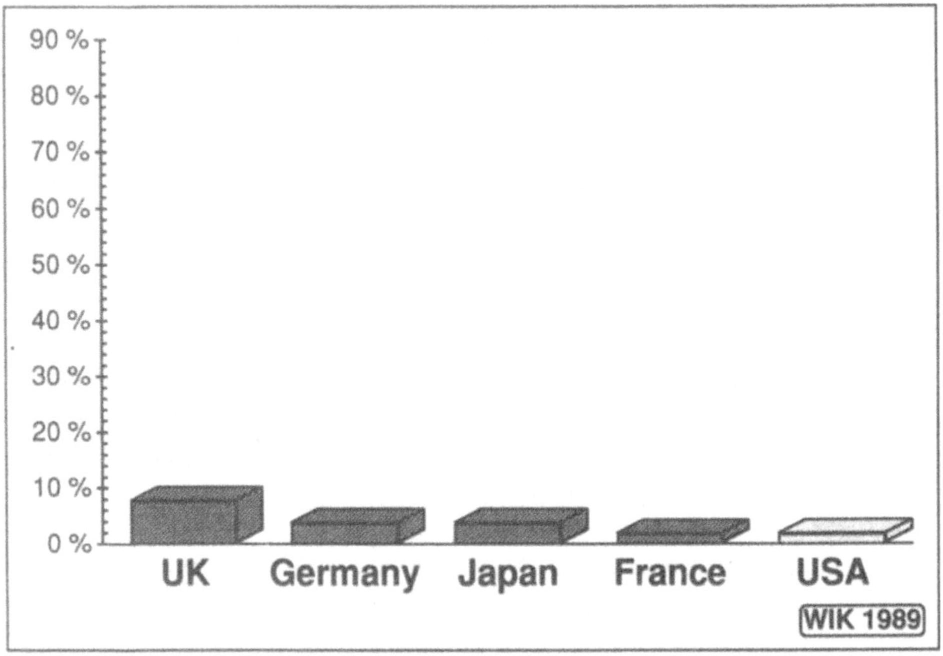

Figure 3.3: R&D for civil telecommunications 1987
Share of manufacturers' R&D expenditure

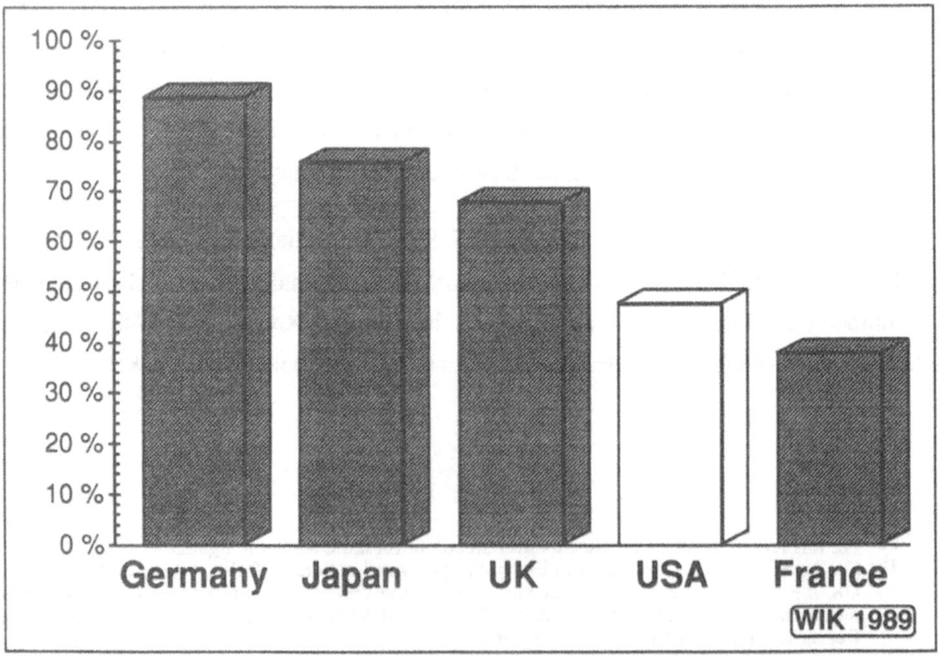

Table 3.2 shows the R&D expenditures of carriers in the five countries. AT&T has by far the largest R&D resources and the highest R&D intensity measured by R&D per turnover. NTT's R&D expenditure is just one third of AT&T's. France Telecom and NTT have the highest R&D intensity within the group of not vertically integrated carriers. The Deutsche Bundespost has the lowest. The Regional Bell Operating Companies (RBOC's) are in the middle.

Table 3.2: Carriers' R&D activities in 1987[b)]

Carrier	Internal and external R&D in Mio. US $[c)]	R&D per turnover in %	R&D personnel
AT&T[a)]	2 600	7.3	26 000
NTT	940	3.8	6 000
RBOC (incl. Bellcore)	910	1.5	6 000
France Telecom	603	4.1	4 300
BT	330	2.1	3 300
GTE	250	1.6	1 800
Deutsche Bundespost	168	1.1	1 800
KDD	25	2.0	190

a) APT included
b) Estimates based on annual reports and interviews
c) National currencies converted into US $ by current purchasing power parities used by the OECD for the field of science

The numbers indicate that AT&T, NTT, France Telecom and the RBOC's are important players in the global R&D activities.

Table 3.3 shows estimates of the R&D expenditures of the ten largest telecommunications equipment manufacturing companies. They all belong to the countries studied in this paper with the exception of Ericsson from Sweden and Northern Telecom from Canada. Many of the companies have diversified business interests outside the telecommunications business. They perform R&D for all their business areas and usually do not publish R&D numbers separately for the telecom sector. Therefore, the numbers are rough estimates.

Table 3.3: R&D expenditures of major manufacturing company groups for civil telecommunications in 1987[a)]

Rank	company group	Telcom-equipment turnover in billion US $	in % of total turn-over	R&D for telecom in million US $[b)]	R&D intensity of the company group in %
1	AT&T (with APT)	11.9[c)]	29	2 600	7.3
2	Alcatel[d)]	8.2	79	800	10.0
3	Siemens[d)]	5.1	24	650	12.5
4	Northern Telecom	4.8	100	588	12.1
5	NEC[d)]	4.1	32	561	10.0
6	Ericsson[d)]	3.3	90	300	9.9
7	Motorola	3.1	52	270	7.8
8	Bosch[d)]	2.5	23	219	6.0
9	GEC/Plessey (GPT)[d)]	2.3	19	437 (210)	3.8 (12)
10	IBM	2.1	4	n.a.	9.3
11	Fujitsu[d)]	1.6	16	≈ 234	8.0
12	Philips	1.5	6	100	7.0

a) Estimates
b) In some cases rough estimations based on various informations from the companies
c) "Sale of products"
d) National currencies converted into US $ by current purchasing power parities used by the OECD for the field of science
n.a. = not available

AT&T has by far the largest R&D budget followed by Alcatel, Siemens, Northern Telecom and NEC.

4. International Comparison of Patents

R&D is a process with different stages and R&D expenditure is only the input side of it. More R&D input does not necessarily mean better R&D results. Better R&D results of a company or a country does not necessarily mean better competitiveness and higher market share. But in high technology sectors such as telecommunications, R&D expenditure and R&D results seem to be a necessary condition for future market success. Patents can be interpreted as an intermediate output of R&D which is especially related to development activities and therefore relatively close to the market. An increasing patent activity in a certain area of technology may be interpreted as an indicator for future market activities and success.

Therefore the Fraunhofer Institut for Systemtechnik and Innovationsforschung (ISI) has analysed patent activities in the telecommunications sector and subfields of it. This analysis was based on patent applications at the US and the German patent office from 1975 to 1986. The assumption is that the situations on the US and the German markets are representative of the international situation. To exclude patents of minor quality the analysis was limited to patent applications for foreign markets because those patents involve higher costs for the applying company. For international comparisons ISI has developed an indicator for the "relative technological position" (RTP) of countries.[17] It is compiled as an index which is set equal to one for Germany in 1975. Patents can give protection for more than one year. Therefore it is sensible to cumulate patents for a number of years to indicate the "patent potential" (PP) which has been accumulated by a country. Because the percentage of patents which are not prolonged increases drastically after five years ISI has accumulated the RTPs for five years. The patent potential PP5 shows the technological potential of a country as an index.[18]

In the following, the PP5 indicator is used to compare patent activities in the USA, Japan, Germany, France and the UK in the telecommunications sector and in certain subfields of it.

Figure 4.1 shows the development of the technological potential for the whole telecommunications sector from 1979 to 1986. We observe moderate increases of the patent activities in the European countries, larger increases in the USA. But the most striking feature is the dynamic development of Japanese patent activities. The Japanese patent potential overtook that of the USA in 1980. Its increase leveled off in the middle of the eighties, but by this time it was about 40 % higher than that of the USA. To judge by this indicator, Japan has thus taken the lead in telecommunications development activities and has accumulated the largest stock of proprietary knowledge in this field of technology. Japan is far ahead of the USA. Germany has the lead in Europe, but the

17 For definitions see Schmoch, 1988, p 56ff

18 Grupp, H., Schnöring, Th. (ed.), Forschung und Entwicklung im Telekommunikationssektor - internationaler Vergleich mit zehn Ländern, chapter 0 (forthcoming)

European countries are far behind the USA and Japan.

Figure 4.1: Technological potential in telecommunications 1979 - 1986

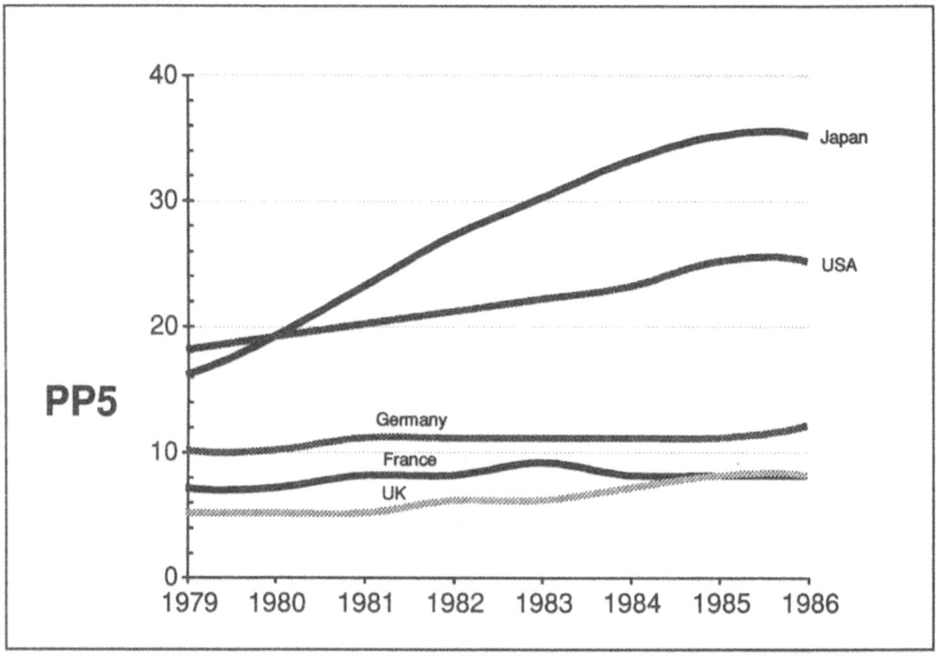

Japanese patent activities have been particularly dynamic in certain subfields of telecommunications: optical communications (optical cables, optical semiconductor elements etc.), image communication (facsimile, picture telphone etc.) and general picture transmission (TV-systems, HDTV etc.). Their technological world leadership position seems to be strong in these areas (Figure 4.2).

Figure 4.2: Technological potential 1979 - 1986

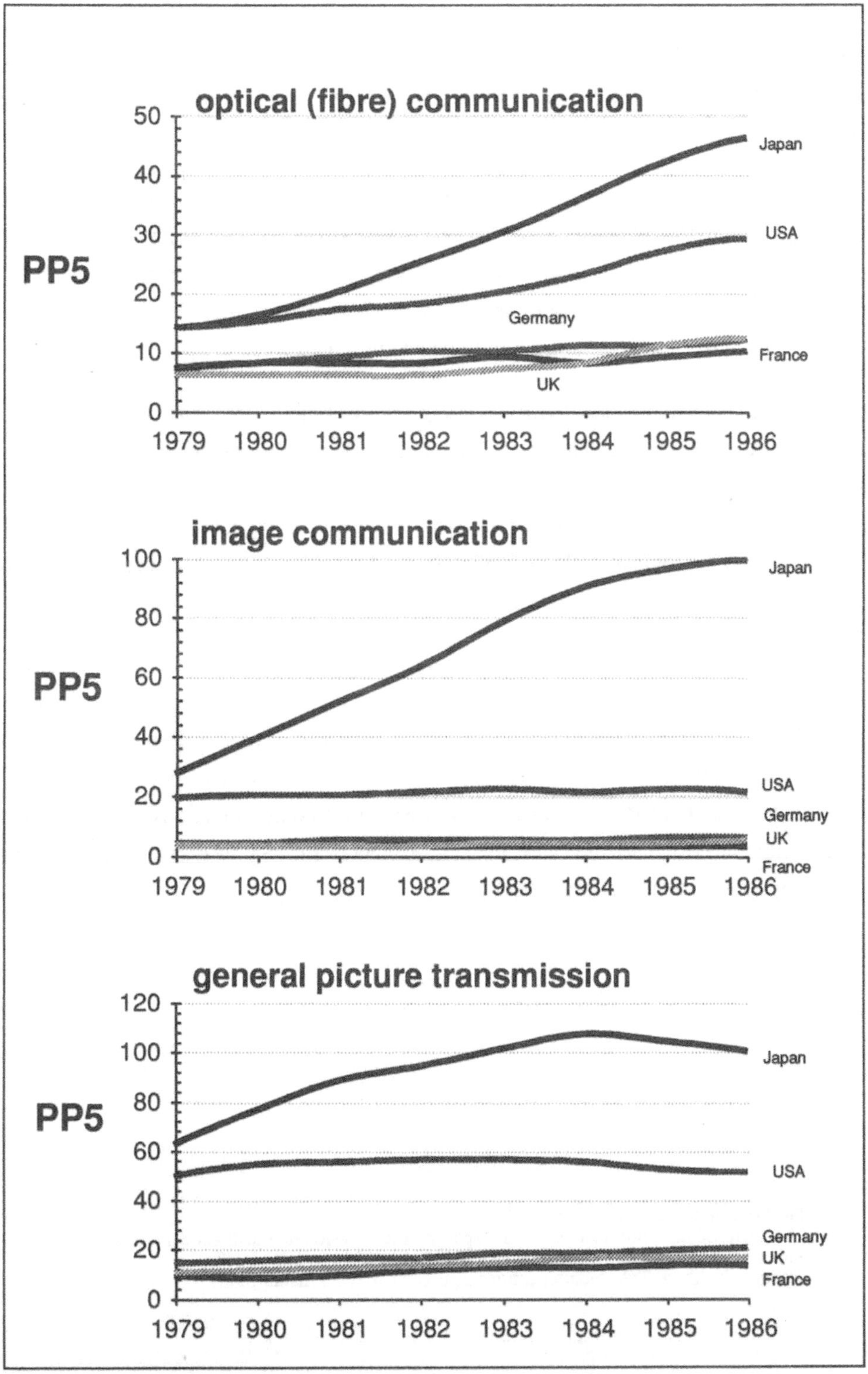

This impression is consistent with other findings: A study of the US National Science Foundation from 1988 revealed an increasing lead of Japan for R&D in fiber optics compared to the USA.[19] Japanese manufacturers have a very strong lead in the global facsimile market. At present, facsimile equipment is the major export product of the Japanese telecommunications equipment industry.[20] A third indication is that Japanese firms (NEC and Fujitsu) have captured almost one third of the US optical fiber electronics market.[21]

The USA still has the lead in other subfields of telecommunications. One is the high frequency telemetry field which is closely related to the military sector. But there are other sectors as well: grid based (wire) communications, radio technique and digital transmission (Figure 4.3). Japan has been catching up in digital transmission and Germany has lost ground in grid based (wire) communications.

Overall, the analysis of patent activities has one message: Japan has invested a lot of resources in building up technological potentials in the telecommunications sector. It has acquired a very large stock of proprietary knowledge in all areas of telecommunications and has taken the leading positions (from the USA) in a number of economically interesting technological fields. Corresponding market successes can already be observed in some markets closely associated with those technological fields.

There remains an open question: Can Japan, based on this large stock of proprietary knowledge, take the lead in other market segments as well? The answer to this question is partly related to the question of patentability of R&D results. Software and systems engineering know-how can hardly become subject of a patent but both are very important elements of R&D and market success in telecommunications. In this respect, patents have a deficit as indicator for future market success in market segments like central office switches and large PABXs.

19 The Wall Street Journal, 17 Nov. 1988

20 The share of facsimile equipment in Japanese total telecommunications equipment export (wire and radio communications equipment) was 20% in 1985 and 45% in 1988.

21 Salomon Brothers, Telecommunications equipment, the US market, 1987

Figure 4.3: Technological potential 1979 - 1986

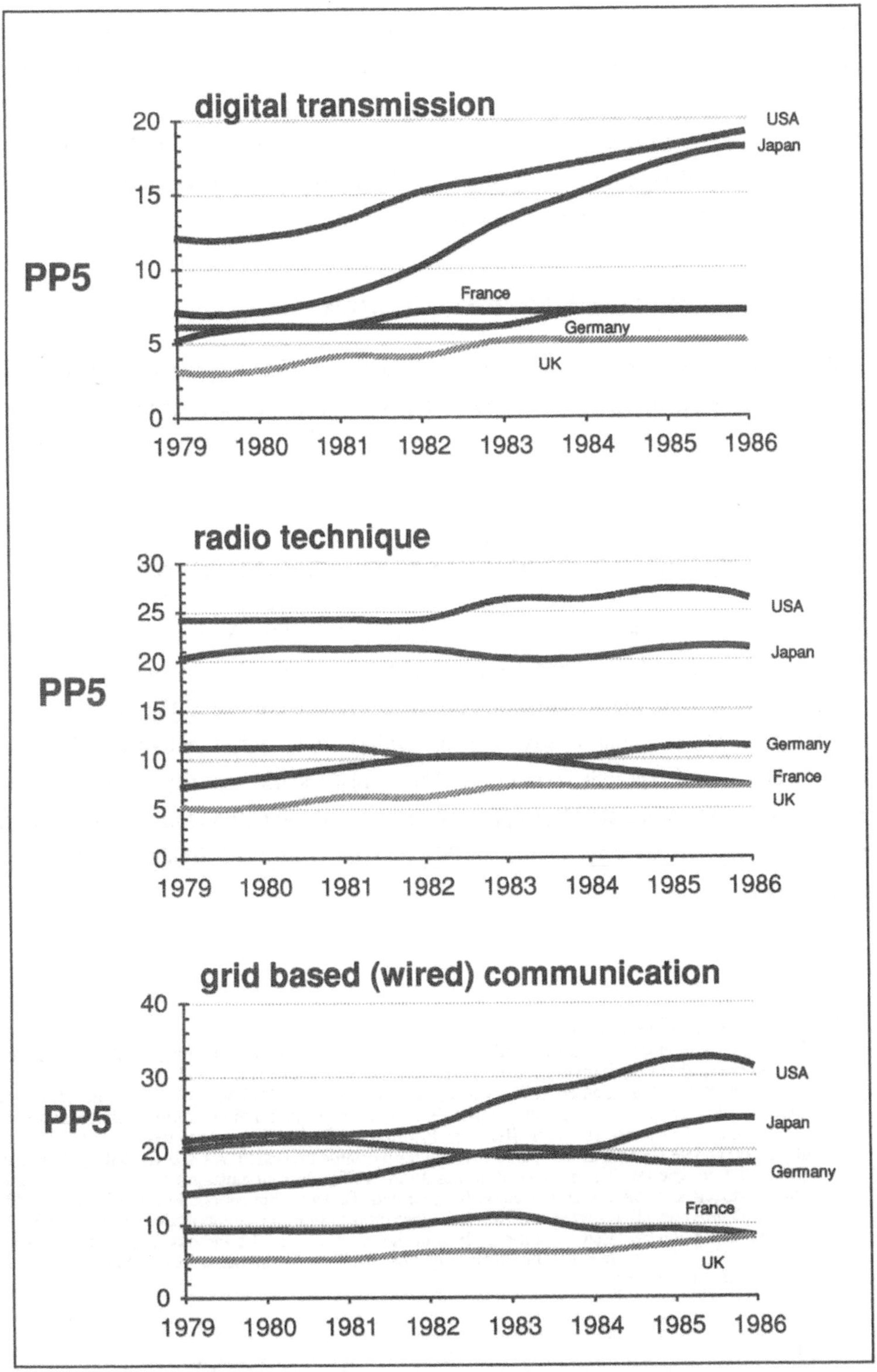

It is often said that Japan has a lag in software development and systems engineering. The US National Science Foundation assesses a lag of Japanese R&D in "telecommunication networks" compared to the USA[22] and Japan's international market position in the switching market is relatively weak up to now.[23] This lag should give some time to European and US competitors in markets for complex telecommunication products. But it is one major objective of Japan's current R&D policy to upgrade the country's software development capabilities. If Japan succeeds in this, then Japanese manufacturers will be strong competitors in all telecommunications market segments based on their large stock of proprietary knowledge.

5. Characteristics of the National R&D Systems

In this chapter, some basic characteristics of the national R&D systems are outlined. The analysis is focused on the general roles of government, carriers, manufacturers, and universities.

5.1 Japan

How can the dynamic and successful R&D activities in Japan be explained?

The Ministry of Trade and Industry (MITI) and the Ministry of Post and Telecommunications (MPT) have a substantial influence on telecommunications R&D. The influence is not so much based on government funding of R&D (Figure 3.1) but on the creation of a complex structure of collaborative R&D activities of NTT, manufacturers, and research institutions. The ministries use a varity of instruments. Recent examples are the R&D program "Telecommunications Frontier Project" (TFP), the Japan Key Technology Center (KTC) and the Advanced Telecommunications Research Institute (ATR).[24]

---

22 The Wall Street Journal, 17 Nov. 1988

23 Switching equipment (central office and PABX) accounts for only 7% of Japans telecommunications equipment exports. This share has been stable since 1985.

24 Japan Key Technology Center (KTC) was founded in 1985 by MITI and MPT. It is an institution which shall promote basic research in Japan. KTC establishes research institutions. One example is the Advanced Telecommunications Research Institute (ATR) which was founded in 1987. ATR has about 200 people R&D staff and performs long term oriented research. ATR is funded by KTC and by private companies. The personnel comes from NTT, telecommunications manufacturers and other companies. Some researches come from abroad. Telecommunications Frontier Project (TFP) is a R&D program MPT started in 1988. It is a program to promote collaborative research of NTT, manufacturers and research institutions. It is designed similar to the famous large scale projects which MITI has created in the early eighties. MPT's Communications Research Lab is coordinating the project.

These programs and projects create collaborative R&D activities of NTT, manufacturers and public research institutions in certain fields of technology for a limited number of years. Projects are often performed in a single lab and the participants delegate R&D personnel. This rotation of staff seems to be an effective mechanism to share existing know-how and to transfer the results of projects back to the companies. The programs generate a sort of common technological background. They have medium and long-term oriented R&D objectives and exclude product development. Product development is done competitively among the manufacturers. From an economic and Western perspective it is astonishing that competing companies are willing to share their R&D resources to such an extent as they do not get proprietary knowledge for their R&D investment.[25]

Since the liberalisation of the telecommunications market and the privatisation of NTT the MPT has developed a stronger commitment to long term oriented R&D and has gained influence on national R&D policy in telecommunications. MPT has enlarged the scope and the resources of its own Communications Research Lab (CRL), it has started TFP and became partner of MITI in KTC and ATR. This may be interpreted as a reaction of the regulatory body to compensate NTT's tendency towards a more market oriented R&D strategy.

NTT, the major carrier in Japan, has large R&D resources (Table 3.2). In the past, NTT had in practice industrial policy obligations which may have resulted in an over-investment of R&D. Its R&D intensity is a lot higher than that of other not vertically integrated carriers. Approximately half of NTT's 6000 R&D staff is working in eleven laboratories. One lab does basic research, the others do applied research and long-term development in certain areas.[26] The other half of the R&D staff is engaged in development activities mainly for software and services.[27] This indicates on the one hand that NTT has large development resources but that NTT is also strongly committed to research and long-term R&D activities.[28] Almost one third of all Japanese scientific publications in telecommunications comes from NTT while the company's share of

25 Some European countries and the European Commission have tried to develop similar programs for "pre-competitive" R&D.

26 The labs are named: Basic Research Lab (approx. 160 people), Applied Electronics Lab (370), Opto-electronics Lab (240), LSI Lab (240), Software Lab (400), Radio Communication Systems Lab (200), Transmission Systems Lab (200), Communication Switching Lab (230), Human Interface Lab (230), Communications and Information Processing Lab (470), Telecommunication Netzworks Lab (200). This gives an indication of the wide range of NTT's research activities.

27 NTT has got a Network Systems Development Center with (approx.) 820 people and a Software Development Center with 1750 People. The rest of the development personnel is in the service divisions.

28 This interpretation seems fairly safe although the definition of what "research" is differs among the countries.

Japanese patents is just 2 %. This is another indicator for NTT's commitment to research and its leading role in the national research community. It remains unclear to what extent the new competitive environment of NTT will limit its willingness and its resources for research in the future. NTT has decentralized some of its development staff to the service divisons.

Japan's manufacturers finance and perform almost three quarters of the national R&D. NEC, Fujitsu, OKI and Hitachi, the so-called Denden-family with traditionally strong relations to NTT, are the leading companies but other well-known electronic companies are engaged in telecommunications R&D as well. It is a charcteristic of most Japanese equipment manufacturers that they all have diversified business interests in the electronics market.

Japan's manufacturers have a strong and broad commitment to long-term oriented R&D. About 5 % of their R&D budget is for basic research and 20 % is for applied research. One third of the scientific publications in telecommunications comes from manufacturers. The publications of the leading companies cover a broad spectrum in telecommunications while the activities of the remaining companies are more focused on particular fields within telecommunications. All companies are engaged in government R&D programs and projects.

The manufacturing sector holds 95 % - 98 % of the patents which indicates that most of the development is done in this sector.

It is not totally clear how NTT and the manufacturers cooperate in the development of complex equipment systems, for example central office switches. NTT's procurement policy says that NTT cooperates with competent manufacturers in the development of equipment which is not available in the market and which needs a lot of special development for NTT's network. In practice this seems to mean that NTT is engaged in joint development of equipment systems with the members of the Denden-family and other companies. This is indicated by many joint patent-ownerships of NTT and manufacturers.[29] It is not known that NTT pays separately for the development activities of the manufacturers. Development costs are included in the product price. This procedure gives NTT's needs a high priority in the development of complex telecommunications systems. This can be a problem for the international

[29] From 1980 to 1986 roughly one quarter of all Japanese NTT patents were joint patent-ownerships with other companies, mostly with one firm of the Denden family. Kobayashi, M. (1987), A mathematical model of collective research, MA Dissertation, Graduate School of Policy Science Saitama University, Urawa 1987

competitiveness of Japanese manufacturers because the systems may not be so well suited for foreign markets with their different needs.[30] But they have a cost advantage because NTT uses substantial own R&D resources for the joint development projects.

Universities do some research in telecommunications but their major role is in education. Two public research institutes outside the university system are important for the government's R&D policy and the whole R&D system: MITI's Electrotechnical Lab (ETL) with 175 people in telecommunications research and MPT's Communications Research Lab with 430 people.

Overall the Japanese R&D system in telecommunications is a complex structure of collaboration of all sectors which realizes a substantial amount of economies of scale and scope. The collaborative approach is particularly strong in research and long-term development, competitive elements are stronger in development activities close to the market. The carrier NTT performs a lot of research and is engaged in joint development with the major manufacturers. The dominant role of NTT may be a problem for export activities on the one hand but it gives Japanese manufacturers a cost advantage on the other. MPT has established itself as a major player in government R&D policy beside MITI. The complex structure of R&D collaboration in Japan can be described shortly as government guided national R&D strategy.

5.2 USA

The Pentagon finances more than one third of the national telecommunications R&D budget (Table 3.1). Its projects are primarily for telecommunications equipment for military use. Spill-over effects to equipment for civil use are likely but they can not be quantified. There is a growing tendency in the USA to use Pentagon R&D programs for industrial policy purposes in the civil sector (SEMATECH and HDTV). For satellite communications, NASA programs are important. Beside that, the US government is not directly involved in the civil telecommunications R&D system.[31]

The carriers AT&T, Regional Bell Operating Companies (RBOC's) and GTE perform half of the civil telecommunications R&D (Table 3.1). AT&T has the largest R&D

30 It may be that this effect is part of the explanation why Japan's international market success in telecommunications equipment has been mainly limited to customer's premises equipment so far. Japans telecommunications equipment exports in 1988 consist to 45 % of fascimile equipment and to 11 % of telephones and telephone application equipment. Switching equipment (central office and PABX) had a constant share of 7-8 % since 1985. The increase in exports of 30 % (1985 to 1988) is almost totally due to the facsimile boom.

31 Government telecommunications policy and the regulation of the industry naturally have indirect effects on the R&D performance of AT&T and the RBOC's.

resources for telecommunications in the world (Table 3.3). Predictions that AT&T would reduce its R&D investment (especially its world-famous research capacity in Bell Labs) after the divestiture have not come true. AT&T's total R&D investment has even increased and the research organization at Bell Labs remained roughly the same size.[32] The RBOC's have created a new research entity, Bell Communications Research (Bellcore). Bellcore is owned by the seven RBOC's and their operating companies, and performs applied research and other technical services for them. It is about 40 % of the size of Bell Labs. The development of Bellcore was promoted by staff transfers from the old Bell System. Recently, some RBOC's started to build up their own R&D capacities besides Bellcore. Overall the research as well as the development expenditures for telecommunications have increased substantially as a consequence of divestiture.[33] The predicted negative impact of divestiture on the US R&D input which was based on the appropriability argument has not occured. The more competitive market environment seem to have stimulated R&D activities.

The RBOC's invest "only" about 1,5 % of their turnover in R&D. In the USA the question has been raised if this is not an underinvestment in R&D.[34] Such a conclusion is mainly based on a comparison with the R&D intensity of manufacturing companies and vertically integrated carriers like AT&T. In absolute numbers, the RBOC's combined R&D budget is the third largest of any carrier worldwide. It has almost the size of NTT's R&D budget which is the largest of any carrier within the group of not vertically integrated carriers. But the RBOC's R&D intensity is relatively low among this group, only the Deutsche Bundespost has a lower one (Table 3.2).

AT&T, the RBOC's and GTE are the most important players in the US R&D system but there are other established manufacturers and a growing number of new manufacturers with R&D resources. The new entrants into the telecommunications market use R&D investment as a strategic weapon and have on average higher R&D intensities than the incumbants.[35] Patents and publications in telecommunications from US sources show that the national R&D activities are widely spread over many entities. AT&T holds the first rank, but with only 11 % of the publications (1986) and 14 % of the patents (US patent applications in 1984 at the USPTO).[36] This indicates a relatively wide-spread

32 Noll (1987) p. 170 f.

33 Noll found an increase of nearly 60% for the combined R&D expenditure of AT&T and Bellcore in 1985 over the pre-divestiture level of 1982. The research organization of Bell Labs had 1249 people in 1982. In 1985 it had 1191 people and Bellcore's applied research group had 498 people. Combined they had 1689 research people. Noll (1987) p. 178

34 See for example Noll (1987) p. 177

35 Dowling, M.J. (1988), Strategic investments in innovation: The telecommunications equipment industry, 1975 - 1986, University of Texas at Austin, p. 49 f.

36 RCA Corporation, ITT, Motorola, GTE, General Electric Company, Zenith, IBM and

distribution of development activities in the USA.

The USA has a diversified infrastructure of public research institutes in telecommunications. Universities and research institutions such as Rand Corporation do research in telecommunications. From an international perspective, their work is relative closely related to industry R&D activities.[37]

Overall, the US R&D system for civil telecommunications has been shaped by its big market and by a relatively large number of carriers and manufacturers which do R&D. From a worldwide perspective, the carriers, especially AT&T, have large research as well as development resources. The government is not directly involved in civil telecommunications R&D, but indirectly over Pentagon funds for R&D.

5.3 France

In France the defence department finances a substantial share of the national telecommuncations R&D budget which has potential spill-over effects on civil telecommunications. The French government uses the national carrier France Telecom for its industrial policy. France Telecom has to finance government R&D programs for the whole information technology sector. The financial burden from this has been sometimes even higher than France Telecom's R&D expenditure for telecommunications. French governments have a long tradition in industrial policy and in telecommunications "industrial policy has been applied in a consistent way according to the motto: promote the national champion(s)"[38].

The French R&D system in civil telecommunications has a relatively simple structure. France Telecom is the dominant entity, with the largest R&D budget of any European carrier and a R&D intensity which is higher than NTT's (Table 3.2). France Telecom's "Centre National d'Etudes des Telecommunications" (CNET) performs most of the national research in telecommunications and finances some external research. CNET also performs long-term oriented development activities for telecommunications systems. Beside that, France Telecom's Department for Industrial and International Affairs (DAII) finances about 45 % of the development in the manufacturing sector directly by development contracts. CNET specifies and controls the contents of the

---

Westinghouse Electr. Company had the most patents after AT&T (US patent applications in 1984 at the USPTO).

37 In 1986, 29 % of the US publications in telecommuncations came from US universities, 6 % from research institutions. In 1984, the research institutions hold 10 % of US patents in the US market, in the German market their share has been only 1 % in 1985.

38 Dang Nguyen, G. (1988), Telekommunications in France, in: Foreman-Peck, J., Müller, J. (eds.), European Telecommunication Organisations, Baden-Baden, p. 138

development contracts while the business aspects of these contracts are handled by the procurement agency. The manufacturing of the equipment is procured separately according to detailed specifications of France Telecom. The specifications are based upon CNET's R&D activities and the results of the development contracts.

Although France Telecom is financing directly most of the development it holds less than 10 % of the French patents while the manufacturers hold almost 80 %. This indicates that the proprietary rights gained in the development activities go predominantly to the manufacturers.

As mentioned above, French governments are using France Telecom's financial resources to finance public R&D-Laboratories and government R&D-programs not only for telecommunications but for the whole information technology. We estimate that roughly 5 % of the French civil telecommunications R&D is financed this way. In any case it is fair to say that there is no other major government R&D funding in this sector besides France Telecom.

The manufacturing sector is highly concentrated with the Alcatel Group as leading company. Alcatel has got by far the largest R&D expenditure in civil telecommunications.[39] Universities and research institutes are of minor importance for civil telecommunications R&D.[40]

The French R&D system realizes a large amount of economies of scale and scope. France Telecom's leading role (even in equipment development) causes a tendency to allocate the national R&D resources dominantly according to the needs of the national carrier. This may be a problem for the international competitiveness of the manufacturers and their equipment.[41] On the other hand they have a cost advantage because the national carrier pays for the development separately.

In the future, the French model may come under pressure from two sides. First, the creation of Alcatel NV due to the aquisition by CGE/Alcatel of the European telecommunications companies of ITT in 1987 has shifted the economic interests of the leading French manufacturer substantially to European markets. In 1987, close to three quarters of Alcatel NV's turnover was outside of France.[42] Alcatel NV has the potential

39 Alcatel's R&D expenditure in telecommunications in France is about twice as much as that of the following four equipment manufacturers (SAT, TRT, Matra and Jeumont-Schneider) together.
40 In 1986, 18 % of the French publications in telecommunications came from Universities and 15 % from research institutes.
41 The French manufacturers have raised this point in the interviews.
42 In 1985 the Alcatel Group had roughly 40% of its turnover in foreign markets. In 1987 Alcatel NV had 70% of its turnover outside of France.

to become a truely international corporation with R&D activites in many European countries. To achieve this, Alcatel will have to orientate its R&D strategy to the needs of all its major customers and not only to the needs of France Telcom. This will be a new situation for France Telecom and may question its leading role in the French R&D system. Second, France Telecom will come under increasing pressure to liberalize its procurement for carrier's equipment. The European Commission has an interest to act this way as well as the governments of the other European countries with a national equipment industry and a different system of R&D funding. From this perspective it is possible that France Telecom will reduce its direct engagement in development activities.

5.4 United Kingdom

The situation in the UK is similar to the French in some respects. The Ministry of Defence (MoD) finances a large share of the telecommunications R&D budget (Table 3.2). MoD gives money to the industry and has its own research institutes. There are indications for spill-over effects but it is not possible to assess their importance.[43] Some British experts have claimed that the large influence of MoD on R&D in the British electronics industry has had a negative impact on the industry's success in civil markets because it has orientated scarce R&D resources towards market niches.

The British government also finances R&D programs in information technology for the civil sector. The programs are supposed to initiate collaborative research activities among industry, research institutes and universities. Some of the projects belong to the telecommunications sector but financially the programs are of minor importance.[44]

Up to the middle of the eighties British Telecom (BT) had a dominant role in the national R&D system for civil telecommunications comparable to that of France Telecom. This has changed to some extent. BT has a large R&D budget and 3300 R&D staff (Table 3.2). The company finances 25 % of the national R&D for civil telecommunications and it carries out 20 % of it. BT and its British Telecom Research Laboratory account for 19 % of the British publications in telecommunications (1986) and 18% of the British patents (British patents applications in 1985 at the DPO/EPO). BT finances research in universities (5 % of BT's R&D expenditure 1987).

---

43 The interviews in the UK have given some examples for spill-over effects of MoD projects in mobile radio communication. See also, Financial Times, August 15, 1989

44 One example is a program called "Joint Optoelectronic Research Systems" with a budget of 64 Mio. $ for the years 1986 to 1989.

For a long time, BT had a strong direct influence on the equipment development in the manufacturing sector. BT had been engaged in joint development programs with the national manufacturers GEC, Plessey and STC and had development contracts with these companies. Problems in the joint development of the central office switch, System X, and the privatization of BT have lead to a gradual withdrawl of BT from such joint development projects. BT has also reduced its development contracts with the national industry.[45] BT's procurement policy has been opened for technological competition. BT has moved to a policy of second sourcing for central office switches, and in general, BT does not specify the equipment it procures to that extent as it used to do.

Recently, BT has changed its internal R&D funding. The new system gives the commercial divisions more influence on R&D programs. One may expect that this will give short-term and medium-term objectives and telecommunications services a higher priority in the R&D programs.

The structure of the British manufacturing sector is changing rapidly. It has been highly concentrated with GEC and Plessey as leading companies. Both have merged their telecommunications divisions into one company, GEC Plessey Telecommunications (GPT), in 1988. GPT is clearly dominating the British equipment market.[46] At present, Plessey is subject to a hostile take-over bid from GEC and Siemens. GEC and Plessey sell most of their telecommunications equipment in the UK.[47] Due to the joint development programs with BT, their R&D activities have been mostly orientated towards BT and the home market. With the problems of the System X development and BT's (consequent) policy change, GPT seems to have lost the international technological race in the central office switch market. BT has suggested to the Monopolies and Mergers Commission that without GPT forming an alliance with another manufacturer it would be very difficult, if not impossible, to maintain a significant British presence in public switching in the 1990s.[48] This can be taken as an open acknowledgement of the failure of the old joint development approach of BT and the national manufacturers under BT's leadership.

British universities do a considerable amount of research in telecommunications (32 % of the national publications 1986) but their work is not very closely related to industry needs.

45 BT still finances the upgrading of the software of System X by a development contract with GPT.
46 Monopolies and Mergers Commission (1989), The General Electric Company plc, Siemens AG and the Plessey Company plc, p. 33
47 The export ratio of GEC in "telecommunications and business systems" was 6 % in 1988 and Plessey's in "telecommunications" only 3 %, while the export ratio of Siemens in "telecommunications networks and security systems" was 23 % for example.
48 Monopolies and Mergers Commission (1989), p. 34

5.5 Germany

In the Federal Republic of Germany, the defence department is not an important financier of R&D. Government R&D policy is done by the Federal Ministry for Research and Technology (BMFT). In telecommunications, the BMFT has funded R&D for satellites as part of programs for space technology, mainly in collaboration with France, and the BMFT has also programs for R&D programms in telecommunications. One major objective of the programs was to secure the international competitiveness of the German industry. The BMFT mainly funded R&D for optical communications technology, switched-broadband communications, the introduction of a very sophisticated data communication network for the science community and R&D for high definition television. The funds for these programs were of minor importance compared to the national telecommunications R&D budget.[49] Half of it went to public research institutes and universities, the other half to the industry. The BMFT activities have helped medium-term and long-term R&D activites in the industry to some extent and they had a positive impact on the innovation activity of the Deutsche Bundespost, especially at the end of the seventies.[50]

The most striking feature of the German R&D system is the insigificant role of the national carrier Deutsche Bundespost (DBP). The DBP has a research institute with roughly 340 people.[51] Its main objective is to provide the carrier with an independent judgement of technological developments, and it is involved in international standardization activities. The DBP has approximately 1500 people engaged in development activities. These people are not involved in the development of equipment but in development activities for new networks and services. Overall the DBP performs only 5 % of the national R&D budget (Figure 3.2). It finances some external research mainly at universities and it procures equipment for pilot applications. Therefore, the DBP finances a higher share of the national R&D budget (7 %) than it performs (5 %) but it is still quite low compared to that of other countries (Figure 3.3).

Up to the end of the seventies the DBP had relative close and stable relations with a small group of national manufacturers. It followed a strict policy of standardization which meant that no more than one system or one typ of equipment for any given

49 In the eighties the BMFT spent on average 40 mio. $ per year but since 1985 the ministry has reduced its engagement drastically to roughly 24 mio. $. The whole national R&D budget for telecommunications was 2.5 billion $ in 1987 (Table 3.1).

50 Schnöring, Th. (1989), FuE-Förderung des Bundesministeriums Forschung und Technologie im Telekommunikationsbereich, Wissenschaftliches Institut für Kommunikationsdienste, Diskussionsbeiträge zur Telekommunikationsforschung Nr. 51, Bad Honnef.

51 The staff of the institute has hardly increased during the last 20 years.

application was procured. This limited the amount of technological competition and gave the largest manufacturer Siemens a leading position in the R&D process. In those days the DBP was relative closely involved in the definition of new systems. During the seventies this system failed in the development of a new generation of public switching equipment (EWS) which beside other factors provided an incentive to change the old system. The DBP introduced R&D rivalry among the manufacturers and restrained further direct involvement in detailed technical specification of new systems. The new procurement systems give an incentive to do specific R&D for the needs of the DBP by restricting the market for a number of years to those manufacturers which survive the first two stages of a procurement process. In the first stage, the vendors present pilot equipment which is tested by the DBP and the second stage is a price competition among the successful vendors for a certain amount of equipment.

The DBP pays for R&D costs of the systems in the course of the actual procurement of the equipment. R&D costs are part of the price calculation of the vendors. Therefore, the amount of money the DBP pays for R&D depends on the competitive situation during the first years when a new system is introduced and the threat of potential competition thereafter.[52]

In 1979, when the DBP introduced this new procurement procedure for digital public switching equipment, the participation was limited to vendors who would develop and produce the switches predominantly in Germany. This has changed. The procurement process for digital access cross-connector systems which started in 1988 was open to any vendor, and foreign manufacturers are taking part.

In such a system it is not astonishing that the manufacturers perform and finance almost all the R&D activities (Figure 3.2). Siemens is the leading German telecommunications manufacturer with large R&D resources (Table 3.3) but there are a number of other manufacturers with substantial R&D capacities. Standard Electric Lorenz (SEL) and Philips Kommunikations Industrie (PKI) are part of large multinational telecommunications and electronic companies, Alcatel and Philips, with large international R&D resources. The Bosch Group is establishing itself in telecommunications and AEG has a long tradition in the market.[53] The number of

52 The procurement systems are different, respectively, for switching system equipment, transmission equipment, etc. but the basic principles are the same. They were developed in 1979 for the procurement of digital switches. Siemens, Standard Electric Lorenz (SEL, German company of ITT, now Alcatel) and Philips Kommunikations Industrie (PKI) took part in the first phase. PKI failed and the DBP is introducing EWSD (Siemens) and SYSTEM 12 (SEL/ITT). Both systems perform well in international markets.

53 In 1985, Siemens had 24 % of the patents submitted by German companies to the German patent office (DPA), SEL 5 %, Philips subsidiaries 6 %, Bosch 16 % and AEG 6 %. In 1986, these firms'

technological independent manufacturers seems to be relative large compared to other European countries.

Many universities and a number of public research institutes do research in telecommunications (34 % of the national publications in 1986). The BMFT finances a specialized research institute, the Heinrich Hertz Institut, with more than 250 people.

6. Conclusions

Differing national R&D policies in general and distinct regulations and market structures in the telecommunications sector have created R&D systems that look quite different from each other. This in the past was sustainable because of the relatively isolated situation of each national market. Liberalisation of the telecommunications markets, however, brings pressure on the old national divisions of labour in R&D.

In all countries the government is - directly or indirectly - involved in the telecommunications R&D system. In the USA, France and the UK, the governments finance R&D for communications equipment for military use which may have spill-over effects to equipment for civil use. In Japan, Germany and the UK, the government finances R&D programs directly oriented towards civil telecommunications but the amount of money involved is small compared to the national R&D budgets. The Japanese government has a particularly strong influence on civil telecommunications R&D through its government guidance of the R&D activities of carriers and manufacturers. In France, the government has a strong influence by the dominant role of France Telecom and its industrial policy objectives. All these government policies tend to create market distortions.

The role of the carriers differs substantially between the countries. According to our estimations, they perform and finance from 7 % of the civil telecommunications R&D budget in Germany up to 60 % in France. This situation can have a strong impact on the international competitiveness of the manufacturers because they have to bear very different R&D costs. As long as the national equipment markets were closed the different systems had no consequences. Carriers which did not finance R&D directly had to pay for it (indirectly) with the price of the equipment. But the situation changes when the carrier's equipment markets are liberalized. Manufacturers which have to bear R&D expenditure themselves have a disadvantage compared to manufacturers which get all

---

shares of the publications by all German entrants were: 22 % for Siemens, 7 % SEL, 5 % for Philips subsidiaries, 6 % for Bosch and 3 % for AEG.

their R&D expenditure or a substantial part of it directly paid by the carrier. Therefore - ceteris paribus - German manufacturers would loose the competition with French manufacturers in a totally liberalized European equipment market for carrier equipment as long as France Telecom holds on to its R&D system.

Of course, reality is more complicated than such simple models and so are predictions of the outcome of a total liberalization of the carrier's equipment market in Europe. Market success depends not only on R&D funding but on R&D efficiency, production, marketing and service as well, and considerable entry barriers for telecommunications systems (like R&D for the adoption of the systems, marketing facilities and a service network) limit the amount of international competition anyway. But the observed differences of the national R&D systems are certainly a problem for the European Commission in their attempt to create a liberalized European carrier's equipment market. We would expect that at least some harmonization of R&D financing between Germany and France will be a precondition for any further liberalization of the carrier's equipment market in the European Community.

From a more general perspective we have two questions. First, the question of advantages and disadvantages of vertical integration for R&D in telecommunications. There are theoretical arguments for both sides.[54] Empirically, we observe vertical integration or quasi vertical integration in most countries but there has been no market test of this structure so far. With the liberalization of the telecom markets the situation has changed to some extent. Divestiture has separated the RBOC's from the manufacturing sector. In Europe, BT and the Deutsche Bundespost have loosened their relation with the national manufacturers in R&D and many carriers started a policy of second sourcing from foreign manufacturers. In principle, this development opens market opportunities for foreign manufacturers but there is not enough empirical evidence to answer the first question.

The second question concerns the competitive situation of not-vertically-integrated manufacturers when their vertically (or quasi vertically) integrated competitors have a monopoly or dominant market position on the service side? In such a situation, the "service division" can cross-subsidize the "manufacturing division" by overinvesting in R&D. But it is difficult to know what the "service division" should spend on R&D under normal market conditions and what is an overinvestment. This shows that the liberalization of the service market and the R&D policy of the dominant carriers are closely related to the liberalization of the carrier's equipment market. The different

[54] See chapter 2.

national approaches on the service market and the distinct R&D policies of carriers (and governments) are obstacles to a liberalization of the carrier's equipment market.

The most interesting and perhaps most challenging country of our comparison is Japan. The Japanese R&D system with its complex structure of collaboration and competition of all sectors under the guidance of MPT, MITI and NTT differs substantially from that of all other countries. The system realizes substantial economies of scale and scope in the R&D process at the national level. During the last ten years, Japan has invested a great amount of resources in R&D for telecommunications. It has acquired a large stock of proprietary knowledge in telecommunications and has taken the leading position from the USA in a number of fields which are important for the future development. If the country succeeds in its attempt to upgrade its software and systems engineering capabilities it is quite likely that Japanese companies will become strong competitors to US and European companies in all telecommunications equipment market segments, not only in the customer's premises equipment market.

References

Dang Nguyen, G. (1988), "Telecommunications in France", in Foreman-Peck, J., Müller, J. (eds.), *European Telecommunication Organisations*, Baden-Baden, Nomos Verlag

Dowling, M.J. (1988), "Strategic investments in innovation: The telecommunications equipment industry, 1975 - 1986", University of Texas at Austin

Grupp, H. (eds.) (1986), *Problems of measuring technological change*, Cologne

Grupp, H., Schnöring Th. (eds.) (to be published 1990), *Forschung und Entwicklung im Telekommunikationssektor - internationaler Vergleich mit zehn Ländern*, Heidelberg et al, Springer-Verlag

Kobayashi, M., (1987), *A mathematical model of collective research*, MA Dissertation, Graduate School of Policy Science, Saitama University, Urawa

Monopolies and Mergers Commission (1989), *The General Electric Company plc, Siemens AG and the Plessey Company plc*

Mowery, D.C. (1987), "Forecasts of the impact of divestiture on Bell Telephone Laboratories: An Assessment", *paper prepared for the Telecommunications Policy Research Conference*, Airlie House, Sept. pp. 27-30

Noll, A.M. (1987), "Bell System R&D activities-The impact of divestiture", *Telecommunications Policy*, Vol. 11, No. 2, pp. 161-178

Noll, R.G., Owen, B.M. (1987), "United States v. AT&T, The economic issues", *Studies in Industry Economics, Discussion Paper* No. 140, Stanford University, California,

OECD (eds.) (1981), *The measurement of scientific and technical activities*, Frascati Manual, Paris

Olley, R. E.(1982), "The process of Technology Transfer and Application in Telecommunications: A case study", in Sahal, D. (ed.), *The transfer and utilization of technical knowledge*, Lexington

Salomon Brothers (1987), Telecommunications equipment-the US market.

Schmoch, U., Grupp, H., Mannsbart, W., Schwitalla, B. (1988), *Technikprognosen mit Patentindikatoren*, Cologne, Verlag TUEV Rheinland

Schnöring, Th. (1989), FuE-Förderung des Bundesministeriums Forschung und Technologie im Telekommunikationsbereich, *Diskussionsbeiträge zur Telekommunikationsforschung* Nr. 51, Wissenschaftliches Institut für Kommunikationsdienste, Bad Honnef.

6

Alternative Settlement Procedures in International Telecommunications Service

James H. Alleman

International Center for Telecommunications Management, University at Omaha, Omaha, Nebraska

Paul N. Rappoport

Temple University Philadelphia

Kenneth B. Stanley[1]

Federal Communications Commission, Washington D.C.

The United States has a large and growing balance of payments deficit in international telecommunications services. In 1980, the deficit for these services was $392 million. The most recent figure is for 1988 when the deficit was over $2 billion.

The deficit is growing at an annual rate of 22 percent. It is projected to surpass $2.5 billion in 1989. International message telephone service (IMTS) accounts for most of the deficit. Settlements owned to foreign communications administrations for telephone service now ąccount for 75 cents out of each dollar collected from IMTS customers in this country.

The deficit in telecommunications services is rapidly approaching the U.S. trade deficit in telecommunications equipment, which was $2.6 billion in 1988.[2] Whereas growth in

---

1 The views expressed in this paper do not necessarily represent the views of the FCC.

2 From the economic point of view the trade deficit is not 'bad' per se, but the fact that strong evidence exists that most of this is a pure transfer of dollars, with no service in exchange.

the equipment deficit has slowed in recent years, the deficit in telecommunications services continues to increase rapidly, up 19 percent in 1987. Moreover, trade in services has become negative, only recently.

Several factors contribute to the deficit. However, tariffs and rate structures are two of the principal causes. Differences in rates and rate structures have a significant impact on the demand for international telecommunications service, as well as on the direction and magnitude of the traffic flow between the United States and other countries. More telephone calls tend to originate in countries that have lower tariffs and flexible, peak/off-peak rate structures than in countries with higher tariffs and no differential prices. The differences in pricing policies give rise to imbalances in traffic flows which, in turn, produce the mounting deficits in the U.S. international telecommunications sector.

Accounting rates also contribute to the deficit. These rates are the basis for the division of revenue between the administration in the country in which service originates and the administration in the country in which it terminates. Unlike tariff rates, which are matters of national determination, accounting rates are established through negotiations between communications administrations in two countries. Accounting rates, in combination with traffic imbalances, determine the settlement payments from one administration to another. Accounting rate levels, accounting rate structures, and the division of accounting rates each have a significant impact on the settlement payments between communications administrations in different countries.

The U.S. balance of payments deficit in international telecommunications services raises several public policy issues. The disparities among accounting rates, tariff rates, and the cost of service pose a potential for cross subsidies to occur with the transfer of funds between communications administrations in different countries. Since the telecommunications firms supply a range of services that share common costs, this raises the potential for inter-service subsidies. Moreover, there is a possibility for cross subsidy between services into other areas of the telecommunications sector.

In addition to cross subsidy issues, the provision of international telecommunications between developed and less developed countries raises a special set of issues. The International Telegraph and Telephone Consultative Committee (CCITT) has established a set of recommendations to govern the division of revenue between and among communications administrations that provide international telecommunications services. The basic principle is an equal division of accounting revenue. The guidelines provide that "[U]nder this procedure, the accounting revenue from the traffic exchanged in their relationship is divided between the Administrations of the terminal countries, in principle on a 50/50 basis. Proportions other than 50/50 may be used when the facilities made available by each of the Administrations of the terminal countries are not approximately equivalent, or if Administrations reach agreement on a different proportion when, for example, the costs differ greatly." The FCC has adopted a more stringent policy with respect to international settlements. Its policy calls for uniform accounting rates on parallel routes, uniform terms of the sharing of tolls, and uniform settlement rates.

This paper describes alternative policy options for the settlement process. A simulation model is presented which can be used to analyze the policy options. A simulation model can be used to estimate the impact of the policy options to change the division of accounting rates. The impacts of other changes in the settlement procedures can also be estimated.

A brief explanation of the settlement process and data on the U.S. deficit precede the discussion of the policy options and the simulation model.

1. Settlements in Telecommunications Services

Communications administrations that provide international telecommunications services are joint owners of international transmission facilities. A consortium of parties enter into ownership and operating agreements for each cable facility. The parties contribute investment capital to purchase the cable and they share the operating expenses on a contractual basis. The financial arrangements for the international satellite system, INTELSAT, are more complicated than cable systems but the basic principle is the same. Each INTELSAT signatory contributes investment capital to purchase the satellites and they share operating expenses. Ownership quotas and contributions to the operating expenses of the satellite system are based on each signatory's relative use of the system. In each case, the owners of the cable systems and the satellite system share common costs in the facilities.

Domestic distribution equipment is also used to provide international service. The equipment is purchased and maintained by communications administrations in each country. They are responsible for the equipment. Unlike international transmission facilities, the equipment is not jointly owned by communications administrations in different countries. However, there are common costs for domestic and international services associated with this equipment.

Thus, administrations that provide international telecommunications service share costs in some facilities and they incur costs in other equipment to establish communications links between countries. In effect, the communications administration in the country where service originates purchases service from the administration in the terminating country so that messages can be transmitted to their destination. A procedure was created to reimburse administrations for the costs they incur on behalf of their foreign correspondents.

The accounting revenue division procedure is the system that is used in international telecommunications to provide for the division of accounting revenue between communications administrations in different countries. The procedure is designed to insure that an administration is reimbursed for the service it provides to a correspondent. The basis for the procedure is the accounting rate, which is a rate per traffic unit that is negotiated between communications administrations in a given relation and used for the establishment of international accounts. The traffic unit for IMTS is a conversation minute. The accounting rate is shared equally between administrations, a 50/50 division. The administration in the originating country retains one-half of the accounting rate and it owes the other half to the administration in the terminating country. The accounting rate can be denominated in U.S. dollars, special drawing rights (SDRs), or gold francs. Since accounting rates are negotiated between communications administrations, whereas tariff rates are a matter of national consideration, sometimes subject to regulatory review, the two rates virtually always differ.

Settlements between communications administrations in two countries are conducted on the basis of the net traffic that flows between them during a specified period of time. The time may be a month, a quarter, or some other period that is agreeable to both administrations. Separate settlements are conducted for each service that is provided, for example, IMTS, telex, and telegraph. Net traffic between countries, in conjunction with the accounting rate and its division, determine both the direction of the settlement payment and its amount. The financial transaction between administrations depends on the monetary unit for payment that has been agreed to by them. Accounting rates for IMTS between the United States and several countries, for example the Federal Republic of Germany, are denominated in SDRs so accounting revenue is measured in SDRs. More IMTS traffic flows from the United States to West Germany than vice versa. As a result, settlement payments are made to the Deutsche Bundespost by U.S. common carriers. The payments are made in the monetary unit stipulated in the settlement agreement between communications administrations. If the currency used to settle the account is marks, then the net balance is converted from SDRs to marks. The U.S. common carrier purchases marks and pays the Deutsche Bundespost in marks. As the discussion implies, foreign currency exchange rates have an impact on settlement payments. The impact may be significant during periods of international financial turmoil.

In general, more telephone service originates in the United States destined for other countries than terminates in the United States from other countries. The difference in traffic flows gives rise to the U.S. balance of payments deficit which is the difference between the revenue U.S. international common carriers owe to foreign communications administrations for service from the United States and the revenue they are owed by foreign communications administrations for service to the United States. The net traffic flow is large and it is growing faster than U.S. settlement payments. As a result, settlement payments flow from U.S. international common carriers to foreign communications administrations.

Table 1 summarizes settlement payments for the period 1980-1987. The table presents the largest settlement payments for IMTS, total settlement payments for IMTS, total settlement payments for telex service (a surplus), total settlement payments for telegraph

Table 1: Settlement Payments for 1983-1988 (million of dollars)

	1983	1984	1985	1986	1987	1988
IMTS	$941.5	$1158.6	$1090.8	$1377.6	$1635.6	$2027.9
Telex	-7.6	-20.5	-18.3	-10.9	-8.2	-3.2
Telegraph	28.3	21.3	18.3	12.7	10.5	7.9
Total	$962.2	$1159.4	$1090.8	$1379.4	$1637.9	$2032.6

Source: Kenneth B. Stanley, "Trends in the International Telecommunications Industry, 1975-1988" Federal Communications Commission, forthcoming.

service (a deficit), and a grand total for all three services. The data show that settlement payments are large and they are growing rapidly. With the exception of 1985, they increased in each year during the period 1983-1988. The dominance of IMTS is evident. It is also increasing, from 89 percent of the total deficit in 1980 to almost 100 percent since 1984. The rate of gain of the IMTS deficit is illustrated in Figure 1. Evidence suggests that a large portion of the deficit may be a subsidy paid by U.S. international common carriers to foreign communications administrations.

Figure 1: United States International Telephony Service Deficit

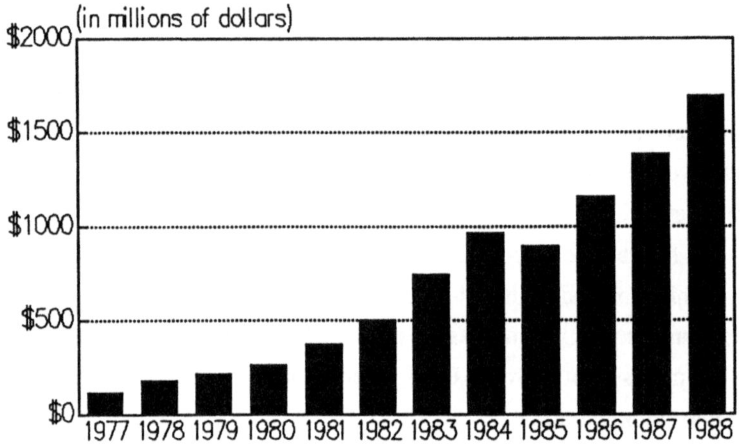

Data excludes Canada and Mexico
Source: FCC, Balance of Payments Deficit

U.S. international common carriers have deficits for telecommunications services with most countries. Many of the largest deficits are with communications administrations in economically developed countries like the Federal Republic of Germany, Japan, Italy, Israel, South Korea, the United Kingdom, France, Taiwan, and Hong Kong. Other deficits are with administrations in less developed countries like Mexico, the Philippines, the Dominican Republic, Pakistan, India, Haiti, and Nigeria. The distribution of the deficit to the top ten countries is shown in Figure 2. The countries included in Figure 2 show that large settlement payments are made to communications administrations in developed countries as well as less developed countries.

Figure 2: Telecommunications Service Deficit-Top Ten Recipient Countries (1988)

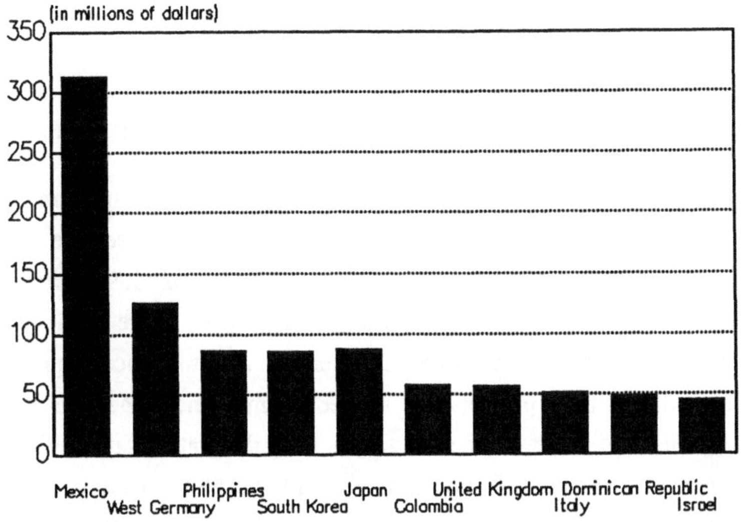

Source: FCC Balance of Payments Deficit.

Alternatives to the existing accounting revenue division procedures range from small changes like revised accounting rates and growth-based accounting rates to more fundamental modifications in the underlying process like "sender keep all" in which the administration in the country where service originates retains all revenue from that service. Proposals to change the procedures have been advanced but aside from revisions in accounting rates the basic procedures have remained intact for some time. The objective of the proposal to change the division of accounting rates is to increase the funds flowing to administrations in less developed countries from developed countries. The justification for the proposal is that the cost of providing service in less developed countries is higher than it is in developed countries. However, there seems to be no cost information to support this allegation or, if it is true, to determine the appropriate revenue sharing arrangement. Even with such information, changing the division may put pressure on communications administrations to initiate other changes. For example, there may be pressure to change tariffs for international service in developed countries if the settlement ratio changes and settlement payments to less developed countries increase even more rapidly. Tariff changes in developed countries may curtail IMTS demand, which has been shown to be price elastic. Instead of leading to an increase in settlement payments to less developed countries, the proposal could have the opposite effect. To experience various alternatives, a simulation model was developed.

2. International Settlements Simulation

This section of the paper develops a policy options simulation model to examine changes in accounting revenue division procedures, including those advanced by communications administrations in some less developed countries. It presents a framework which can be used to evaluate a range of alternative policy options.

A full simulation model requires time to develop, test, refine, and implement. Cost and demand data have to be collected and assessed for international service in the United States and other countries. A large source of usage data is available for service between the United States and other countries. Cost data may be more difficult to obtain. Modeling the demand for international telecommunications service requires extensive work. Outputs from these projects are needed to implement the simulation model. However, a model can be developed and tested using illustrative data until the variables are more accurately estimated. The procedure used to develop such a model is discussed below. Although the estimates are not precise, the authors believe this is a realistic view of the international settlements process.

2.1 Model

A "two" country model is proposed initially: the United States (US) and the rest of the world (ROW). This is not done simply out of provincial motives, but because the availability of data between the U.S. and the ROW is more complete; moreover, a significant portion of the world's international traffic originates or terminates in the U.S. (Because of peculiarities of the data and settlements process, the model will exclude Canada and Mexico).

The simulator requires estimates of price elasticities, cross price elasticities, income elasticities, accounting rates, and accounting rate divisions. The simulation system permits the construction of the relevant matrices governing the interaction of demand, price responsiveness, and demand growth. The simulation system is designed to estimate the impact of numerous assessments, including revisions in accounting rates, modifications in the division of accounting rates, net revenue flows, and settlement payment between countries. Different scenarios can be compared as well as a full assessment of simulated demand.

In summary, the options that are available with the simulation system proposal include:

1) prices for service from the United States and other countries,
2) changes in accounting rates,
3) changes in accounting rate divisions,
4) elasticity parameters, and
5) exchange rates.

The output of the simulation system include:

1) settlement payments between the United States and other countries,
2) revenue changes within countries, and
3) changes in the level of traffic between the United States and other countries,
4) changes in the profits,
5) changes in the consumer's surplus.

2.2 Costs

The lack of solid empirical support for change of the accounting rate division demonstrates the need for reliable, forward looking cost of service estimates that can be used to determine accounting rates and their division for different services between different countries. In telecommunications policy and practice, cost has a variety of interpretations and definitions. In order to estimate the impact of the options and to provide useful policy guidance, cost must be defined in a particular way.

To assess proposals calling for changes in the accounting revenue division procedure based on cost, a cost study methodology should fulfill four requirements. The study: (1) should be forward looking, (2) the measure of costs should be comprehensive, (3) a priori allocation to particular services should be avoided, and (4) it should be flexible and capable of replication.[3] Such studies have not been completed, however. For lack of such studies, the model will use the fully distributed cost estimates developed by AT&T cost of \$.35 per minute for international traffic. This includes both incoming and outgoing traffic for 1985. The model uses the number for the US and ROW. The model is not "inflated" to 1988 because (1) fully distributed cost overestimates the economic cost, as is well known and (2) technology in this area should have maintained or even reduced the cost of the service, e.g., with the recent introduction of fiber optic cable

[3] Goetz, Michael L., "Cost Allocation Techniques and Pricing Alternatives: Crossing the Great Divide," in Alleman, James H. and Richard M. Emmerson, (eds.), Perspective on the Telephone Industry: The Challenge for the Future, Ballinger Publishing Company, 1989.

systems and other technological advances.

2.3 Revenue and Data Analysis

The accounting revenue procedure for IMTS is a difficult process to model. Two factors contribute to the complexity of the process. First, the price of service in a country has an impact on the level of demand in the country as well as the flow of traffic from other countries.[4] Second, monetary exchange rates affect the relative price of telephone service between countries. Exchange rates have varied considerably in recent years and this only contributes to the modeling problem. Moreover, the accounting rate can be denominated in U.S. dollars, SDRs, or gold francs. Financial transactions between administrations can take place in still other monetary units.

The mechanics of the international settlements process are discussed in a simplified model that omits the complications. The model is based on two "countries", US and ROW, which have an accounting rate, r, and an accounting rate division, s_i, i = US,ROW. The settlement rate is defined as $s_i r$.

Let t_i , i = US, ROW, be the traffic from country US and ROW, respectively, to the other country. Then assume $t_i = f(p_{US}, p_{ROW}, X)$, i = US, ROW, where p_{US} and p_{ROW} represent the prices of telephone calls in countries US and ROW, respectively, and X represents a vector of exogenous variables. If F_{US} (F_{ROW}) is the net flow of settlement payments between US and ROW (ROW and US), then

$$F_{US} = s_{US}\ r\ t_{ROW} - s_{ROW}\ r\ t_{US}$$

and $F_{US} = -F_{ROW}$.

The net revenue of a country, R_i, where i = US, ROW is the price of the call, p_i, times the traffic, t_i, originating in the country less the settlement payment (positive or negative). The relationship can be summarized for country US as

$$R_{US} = p_{US}\ t_{US} - s_{ROW}\ r\ t_{US} + s_{US}\ r\ t_{ROW}$$

$$= (p_{US} - s_{ROW}\ r)t_{US} + s_{US}\ r\ t_{ROW}.$$

Similarly for ROW.

4 See Applebe, T.W., A.C. Larson,and D.E. Lehman, "Unilateral Pricing of Telecommunications Traffic," Proceedings of Forum 87, (forthcoming).

2.4 The Demand For International Telecommunications Services

Settlement payments between the United States and other countries depend on the factors that affect the demand for service. The factors include intensive factors such as the responsiveness to price changes and the availability of certain technology, which have an impact on changes in demand in the short run, and extensive factors such as aggregate economic growth, population expansion, and the level of trade and travel. Several studies of international telecommunications demand were conducted in the 1970s.[5] Despite the imposing data problems, these studies, taken together, strongly suggest that overall demand for international telephone service was price elastic; that it was becoming more elastic with time; and that demand for service originating in the United States was growing more rapidly than the demand for service terminating in the United States. A more recent study draws a similar conclusion about the price elasticity of demand for service originating in the United States, although the results are less conclusive for service terminating in the United States.[6]

Most demand studies indicate that IMTS is responsive to price changes. Certainly this has been shown to be the case for service between the United States and many other countries. Recent studies of the demand for international telephone service support these conclusions. Improvements in service quality and more widespread availability of service in less developed countries should only add to the price effects in developed countries. Revenues, therefore, should rise. Moreover, net revenue should rise as indivisibilities are spread over increased levels of output and generate investment funds to upgrade the international telecommunications network.

The changes that have taken place in recent years and the shortage of up-to-date studies demonstrate the importance of modelling and estimating the demand for international telecommunications service. Current estimates of the most significant determinants of

5 The articles include Khadem, R., "An Econometric Forecasting Model of the Demand for International Voice Telecommunications from Canada," Teleglobe Canada, October 1976, Kumins, L., "An Estimate of Price Elasticity of Demand for International Long-Distance Telephony," Telecommunications Journal, June 1, 1967, Lago, A. M., "Demand Forecasting Models of International Telecommunications and Their Policy Implications," Journal of Industrial Economics, November, 1970, Rappoport, Paul N., "A Forecasting Model of MTS Demand between the United States and the United Kingdom," Federal Communications Commission, February, 1979, and Yatrakis, P. G., "Determinants of the Demand for International Telecommunications," Telecommunications Journal, December, 1972, Craver, R. F., "International Calling at Bargain Rates," Telephony, December 15, 1975, Craver, R. F., "A Look into a Crystal Ball at the Future of International Calling," Telephony, August 2, 1976, and Craver, R. F., "An Estimate of the Price Elasticity of Demand for International Telecommunications," Telecommunications Journal, November, 1976. A summary of these articles appears in Taylor, Lester D., Telecommunications Demand: A Survey and Critique, Ballinger Publishing Co., Cambridge, 1980. Also see Rappoport, Paul N., "Forecasting the Demand for International Message Telephone Service," March 1977.

6 See Acton, Jan Paul and Ingo Vogelsang, "Telephone Demand over the Atlantic: Evidence from Country-Pair Data, June 15,1988 (mimeo).

demand are essential to addressing the balance of payments situation. In the 1970s the primary concern was to forecast growth in demand in order to plan the requirements for international transmission facilities. More recently, the focus has shifted because the new issues have emerged. The current concern must be directed to identifying and estimating a more rigorous modelling framework that provides answers to questions concerning settlement issues. Appendix B outlines the proposed method for estimating demand for international voice traffic. In lieu of comprehensive estimates of demands, the model uses point price elasticity estimates of -.8 and -1.2, for the US and ROW, respectively. In the absence of recent elasticity estimates, it was assumed the US elasticity was -.8 and the ROW was -1.2.

Linear demand curves were then constructed from the 1988 quantities and derived U.S. price of $1.18 per minute and the ROW price, which was assumed to be twice that of the U.S. price.

3. Results

Five scenarios are examined:

- Price Reductions in ROW
- Forty/Sixty Accounting Rate Split
- Decrease in the Accounting Rate
- Shift in ROW Demand
- Decrease in All Prices.

Each of the scenarios are examined in turn and their salient points elaborated. The graphic results are reported for the change in profit and the change in consumers' surplus. Their sum is reported as the net benefit for each "country". The total benefit is reported as the sum of the net benefits in each "country".[7]

The complete simulation results and the base case (1988) are reported in Appendix A. (A run time version of the simulation is available upon request from the principle author). One result is rather startling--the magnitude of the profit from this service, both in the US and ROW. And this is "pure economic" profit in the service as defined by the economists, i.e. *after* accounting for a normal return on investments! The level of profit

7 Consumer's surplus has been debated in the literature as a tool for evaluation on benefits. Willig's, R.D. (1976) "Consumer's Surplus Without Apology", *American Economic Review*, 66:589-597, seminal article should have ended this bickering; however, the redistribution question certainly is raised when country comparisons are made.

may not be precise, but the order of magnitude may be well correct, considering that many countries earn up to 75 percent of the revenues from international services.[8]

3.1 Scenario 1: Price Reductions in Rest of World

Scenario 1 is a reduction in the ROW's price by ten percent. The rationale for this scenario is that the prices in the ROW are generally conceded to be relatively high, certainly in comparison to U.S. prices. Moreover, this change illustrates the interrelationship between actions in the rest of the world and the US.

Profits increase in the US because traffic in the ROW has increased, and hence the difference (net traffic) between the countries decreases and the net settlement is less. The large increase in consumers' surplus in the ROW is for the usual reason, i.e. there has been a price reduction. Although not illustrated in the Figure, the collection revenue in the ROW has increased. (See Figure 3)

Figure 3: Benefits Scenario 1

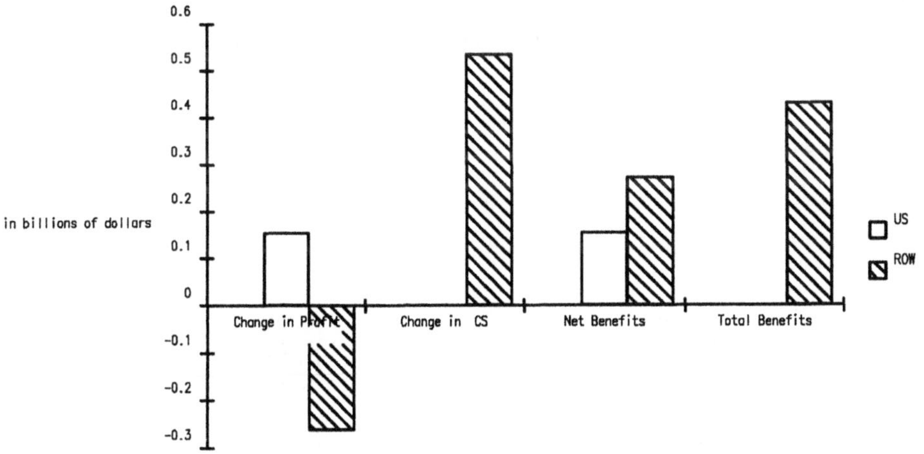

If "hard currency" is an issue, the amount of hard currency lost is indicated by the change in the profit of the US carriers. That is to say, if the ROW only represented non-convertible currency countries, the gain in US profits represents a loss in convertible currencies. This can easily explain the reluctance of developing countries to adopt this

---

8 See Alleman, James H., and Stanley, Kenneth E., Proceedings of the First International Telecommunications Futures Symposium, (forthcoming), Omaha, NE. See also Regulatory Policies and Telecommunications, Notice of Inquiry and Proposed Rule Making, FCC Docket No. 86-494, 2 FCC Record 1022, January 30, 1987, Report and Order and Supplemental Notice of Inquiry, March 25, 1988; and Stanley, Kenneth B., "The Balance of Payments in International Telecommunications Services," Federal Communications Commisssion, December 12, 1988.

type of policy--not only are profits for the PTT decreased, a significant portion of the lost profits are in the form of hard currency.

Abstracting from the hard currency issue, the ROW unequivocally gains with this change, as would be expected, when monopoly prices move closer to costs. The size of the gain is illustrated by the net benefits and, in this case, represents approximately one third and one sixth of a billion dollars a year for the ROW and US, respectively, or approximately five percent of the billed revenues.

3.2 Scenario 2: Forty/Sixty Accounting Rate Split

Scenario 2 (Figure 4) illustrates the change in the settlements ratio to the forty/sixty split of the accounting rate. An across the board change to this division of revenue shows that it is a simple transfer of profits from one administration to another; in this case, a transfer of two hundred million dollars. No total benefit exists, because it is a simple transfer. No gain accrues to the consumers, since under the assumption of the model, this money is not "shared" with the consumers. To be sure, secondary effects may occur. Rates may have to increase on other services in the US to make up loss of profits, and an increase in ROW profits may be returned to the consumer, although this has hardly been the case in the past.

Figure 4: Benefits Scenario 2

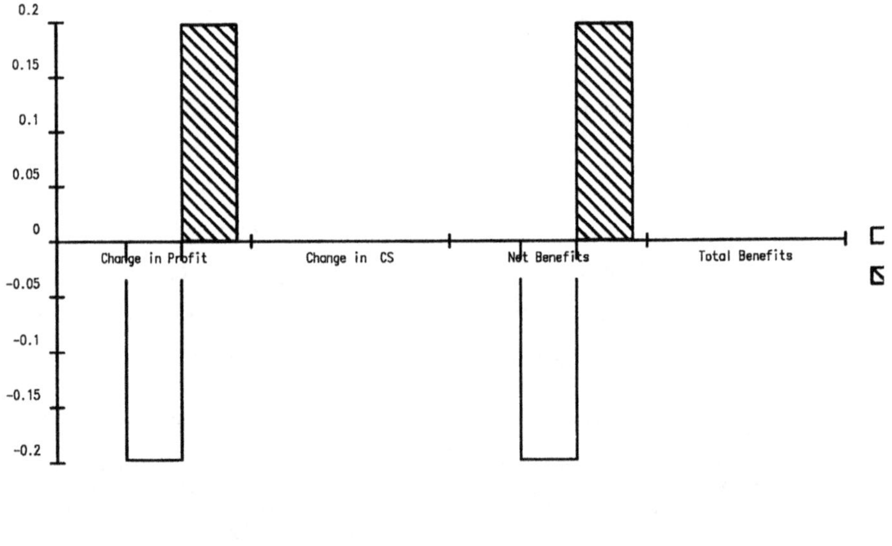

3.3 Scenario 3: Decrease in the Accounting Rate

Scenario 3 (Figure 5) is the reverse of Scenario 2. In this case a lowering of the accounting rate allows the US carriers to retain more of their revenue at the expense of the ROW. Once again, the consumers do not benefit from these transactions. Scenarios 2 and 3 suggest a combination of these two approaches as a policy option, which addresses the twin requirements of an increase in the funds to the developing countries and amelioration of the flow of funds out of the U.S.

Figure 5: Benefits Scenario 3

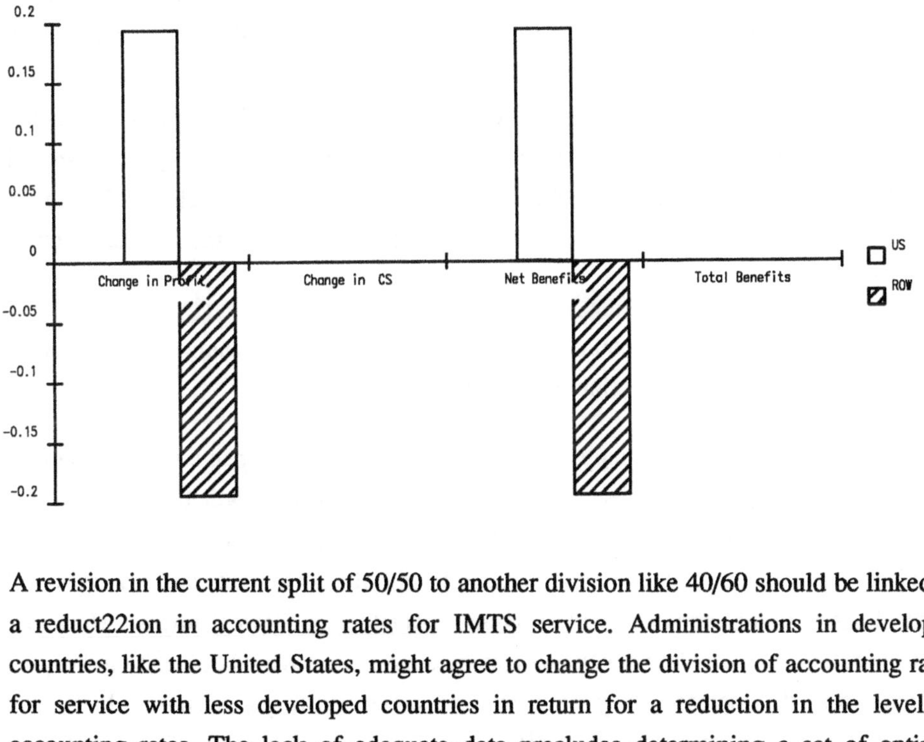

A revision in the current split of 50/50 to another division like 40/60 should be linked to a reduct22ion in accounting rates for IMTS service. Administrations in developed countries, like the United States, might agree to change the division of accounting rates for service with less developed countries in return for a reduction in the level of accounting rates. The lack of adequate data precludes determining a set of optimal accounting rates so actual rates and their division will continue to be set through bilateral negotiations between communications administrations. Initially perhaps a range for the division, like 50/50 to 40/60, should be set as boundaries for bilateral negotiations. A range might also be set for reductions in accounting rate levels, say 10-20 percent per year below current levels. As better information and more rigorous analysis become available, alternative solutions may become attractive and mutually beneficial to communications administrations.

Pressure will mount to take action to ameliorate the flow of funds from the United States to other countries. Undoubtedly communications administrations in developing countries will continue to press the argument that the costs they incur to provide international service are higher than the costs in developed countries and seek relief in the form of modifications in the division of accounting rates so that they receive a larger share of the settlement pie.

The proposal to modify in settlement procedures occurs at a time when settlement payments from U.S. international common carriers are increasing rapidly. The U.S. balance of payments in international telecommunications services may deteriorate further as a result of changes in the division of accounting rates that favor other countries. This will increase the pressure to act. Indeed indications of concern have already been expressed. The FCC, for example, noted its growing concern with the mounting deficit. Perhaps in response to this concern, AT&T, after a period of relative inactivity, renegotiated its settlement agreements with a growing number of communications administrations and many accounting rates have declined. The lower accounting rates, however, have not been sufficient to offset the growing traffic imbalance.

The combination of lower accounting rates and the proposal to change the division of accounting rates suggests an option that could be adopted.

The main issue for communications administrations in less developed countries seems to be a desire to get a bigger piece of the settlement pie. Others will benefit from this outcome if the payments are channeled into communications systems in less developed countries so service is improved and demand is stimulated. Markets and revenues will expand and more services will become available at lower prices. However, administrations in many less developed countries have encountered difficulty in keeping or exploiting the increases they have realized under existing procedures. Inefficient pricing structures, cross subsidies within services, reliance on funds generated in telecommunications service to subsidize other government operations and institutions, and other infirmities have tended to preclude effective utilization of increased revenues from the international settlement process to improve communications networks in less developed countries.

This combination cannot be modeled with a two country model, since the ROW includes developed as well as developing countries. But by interpolating between Scenario 2 and 3 and only applying the forty/sixty split to the developing world, we can see how the twin goals can be addressed.

It seems advisable to attach a set of conditions to changes in the division of accounting rates before concessions are made to current practices. The conditions should be designed to encourage administrations to invest in their communications networks. They are intended to be suggestions; others may be desirable. Care should be taken, however, to concentrate on the most significant conditions so that implementation of the option is not burdened by less important matters and the objective of improving the communications networks is not obscured. Three conditions seem to be particularly important. One is the elimination of cross subsidies from the telecommunications sector to other government sectors. This will insure that revenues generated by the telecommunications sector are invested in capital to improve it. Another condition is to guard against reductions in investment in the country's telecommunications sector as a result of increases in its international telecommunications revenues. Here again, the emphasis is on using telecommunications revenues to improve service. The third condition is to minimize cross subsidies among telecommunications services.

3.4 Scenario 4: Shift in Rest of World's Demand

Scenario 4 (Figure 6) is a gain for all due to the shift in consumers' demand by ten percent. US profits increase because of the reduction in net traffic and the ROW profits increase because its average and margin cost remain constant with the increase in the demand for traffic overtime. The scenario is not unrealistic for some countries. When semi-automatic and automatic dialing is implemented to other countries, dramatic increases in traffic are observed.

Figure 6: Benefits Scenario 4

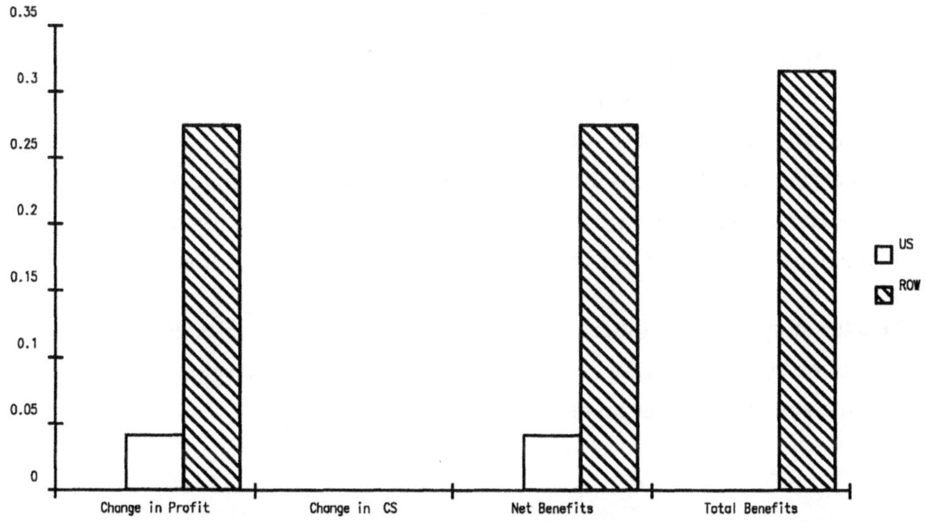

3.5 Scenario 5: Decrease in All Prices

The last scenario (Figure 7) illustrates a price decrease in both "countries". Without the settlement effort, the decrease in price in ROW would lead to an increase in profits because of the elasticity and constant original cost assumptions. But when settlement is included, traffic in ROW increases more than the increase in the US and, thus, the net traffic decrease and the settlement is reduced correspondingly. The profit fall in the US is ameliorated by the reduction in settlements, however. The major gainers in both countries are the consumers. Their total gain is over two billion dollars a year.

Figure 7: Benefits Scenario 5

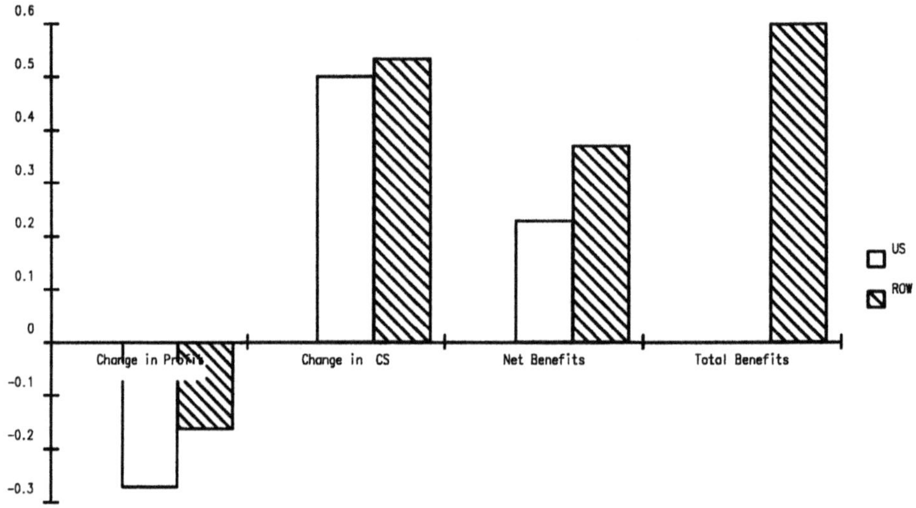

In general, the simulation illustrates the magnitude of the profits which are being generated in the international service, presumptively due to historical monopoly power enjoyed by these countries. With the introduction of competition in the US market, this power is reduced, but it has not affected the institutional structure of the ROW carrier. The old arrangements have remained in place for the most part, and created an asymmetry in the settlements process. The above scenarios illustrate how difficult it will be to achieve consensus on policy alternatives. Large amounts of money will change hands and neither side has any incentive to make a transfer to the other.

The other story of this analysis is the globalization of the economics. Regulatory policies and actions by the telephony do not occur in a vacuum, but have ramifications beyond the borders of the countries involved.

4. Conclusion

There is a large and rapidly growing U.S. balance of payments deficit in international telecommunications services. The size of the deficit and its growth raise several public policy issues. Some steps are being taken to bring the deficit under control but in spite of these efforts it continues to grow. In the midst of this problem, a new issue has emerged that threatens to fuel additional growth in the deficit. Communications administrations in some less developed countries have proposed to change the division of the accounting rates from an equal split to one that is more to their advantage. Instead of receiving an equal share of accounting revenue, this proposal calls for some administrations to get a larger share. The administrations advancing the proposal argue that it is justified because their international telecommunications costs are higher than they are in other countries. There is no empirical support for this claim. U.S. international common carriers can expect their settlement payments to increase even further if the proposal is adopted.

The international settlement procedure is a complicated process. A change in one element of the process has impacts on other parts. An analytical framework is necessary to estimate the impact of a change like the modification in the division of accounting rates that is being proposed. The analysis should not be limited to measuring costs to see if they are different among communications administrations. A more comprehensive analysis should be performed to estimate the impact on revenues, demand for service, and international settlements. The simulation model is a start. The additional data and more accurate estimates of the parameters a full simulation requires should be collected. Such an effort will take time and will require cooperation by communications administrations.

Changes in the accounting revenue division procedures may be warranted. The U.S. balance of payments deficit in international telecommunications services points to the need for more immediate attention. Any change in the division of accounting rates should be linked to reductions in the level of accounting rates. A range should be established that sets bounds for the division of accounting rates. A range of 50/50 to 60/40 is suggested. Conditions need to be attached to changes in the accounting revenue division procedure to insure that increases in settlement payments are used to finance improvements in communications networks.

APPENDIX A

International Settlement: Alternative Policies Simulation

Table A-1: International Settlements: Base Case (1988)

	USA	ROW
Gross Revenues (billions)	$4.45	$4.49
Settlements (billions)	$-1.62	$1.62
Elasticity	-0.8	-1.2
Price	$1.18	$2.36
Quantity (billions of minutes)	3.77	1.90
Cost per minute	$0.35	$0.35
Accounting Rate	$1.69	
Settlement Ratio	0.5	
Effective Rate	$0.85	
Net Revenues (billions)	$2.83	$6.10
Costs (billions)	$1.98	$1.98
Profit (billions)	$0.85	$4.12

Table A-2: Alternative: Ten percent reduction in ROW Price

	USA	ROW
Revenue (traffic) (billions)	$4.45	$4.55
Settlements (billions)	$-1.38	$1.38
Elasticity	-0.8	-1.2
Price	$1.18	$2.12
Quantity (billions of minutes)	3.77	2.14
Cost per minute	$0.35	$0.35
Accounting Rate	$1.69	
Settlement Ratio	0.5	
Effective Rate	$0.85	
Net Revenues (billions)	$3.07	$5.93
Costs (billions)	$2.07	$2.07
Profit (billions)	$1.00	$3.86
Change in Profit	0.15	-0.26
Change in CS	0.00	0.53
Net Benefits	0.15	0.27
Total Benefits		0.43

Table A-3: Alternative: Forty/Sixty Settlement

	USA	ROW
Revenue (traffic) (billions)	$4.45	$4.49
Settlements (billions)	$-1.90	$1.90
Elasticity	-0.8	-1.2
Price	$1.18	$2.36
Quantity (billions of minutes)	3.77	1.90
Cost per minute	$0.35	$0.35
Accounting Rate	$1.69	
Settlement Ratio	0.4	
Effective Rate	$1.02	
Net Revenues (billions)	$2.55	$6.38
Costs (billions)	$1.98	$1.98
Profit (billions)	$0.56	$4.40
Change in Profit	-0.20	0.20
Change in CS	0.00	0.00
Net Benefits	-0.20	0.20
Total Benefits		0.00

Table A-4: Alternative: Decrease in Accounting Rate by Ten Percent

	USA	ROW
Revenue (traffic) (billions)	\$4.45	\$4.49
Settlements (billions)	\$-1.42\$	1.42
Elasticity	-0.8	-1.2
Price	\$1.18	\$2.36
Quantity (billions of minutes)	3.77	1.90
Cost per minute	\$0.35	\$0.35
Accounting Rate	\$1.52	
Settlement Ratio	0.5	
Effective Rate	\$0.76	
Net Revenues (billions)	\$3.02	\$5.91
Costs (billions)	\$1.98	\$1.98
Profit (billions)	\$1.04	\$3.93
Change in Profit	0.19	-0.19
Change in CS	0.00	0.00
Net Benefits	0.19	-0.19
Total Benefits		0.00

Table A-5: Alternative: Shift in ROW Demand Curve

	USA	ROW
Revenue (traffic) (billions)	$4.45	$4.93
Settlements (billions)	$-1.51	$1.51
Elasticity	-0.8	-1.2
Price	$1.18	$2.36
Quantity (billions of minutes)	3.77	2.09
Cost per minute	$0.35	$0.35
Accounting Rate	$1.80	
Settlement Ratio	0.5	
Effective Rate	$0.90	
Net Revenues (billions)	$2.94	$6.44
Costs (billions)	$2.05	$2.05
Profit (billions)	$0.89	$4.39
Change in Profit	0.04	0.27
Change in CS		0.00
Net Benefits	0.04	0.27
Total Benefits		0.32

Table A-6: Alternative: Decrease in Prices by Ten Percent

	USA	ROW
Revenue (traffic) (billions)	$4.34	$4.55
Settlements (billions)	$1.65	$-1.65
Elasticity	-0.8	-1.2
Price	$1.06	$2.12
Quantity (billions of minutes)	4.09	2.14
Cost per minute	$0.35	$0.35
Accounting Rate	$1.69	
Settlement Ratio	0.5	
Effective Rate	$0.85	
Net Revenues (billions)	$5.99	$6.19
Costs (billions)	$2.18	$2.18
Profit (billions)	$3.81	$0.72
Change in Profit	-0.27	-0.16
Change in CS	0.50	0.53
Net Benefits	0.23	0.37
Total Benefits		0.60

APPENDIX B

The following system presents a modelling framework for international telecommunications demand:

$$TD_{A+US} = f(EX, GNP_A, GNP_{US}, V, BT, P_A, P_{US}...) \quad (1)$$

where:

- TD is the total demand for service between country A and the United States,
- EX is the exchange rate between the country's currencies,
- GNP is the measure of aggregate economic activity in each country,
- V is a vector of relevant government policies affecting telecommunications,
- BT is the level (growth) of bilateral trade between country A and the United States.
- P_i is the price of the international call in the country, i = A and US.

Equation (1) describes the extensive factors that determine the growth in demand. We postulate that service originating in the United States is described by the change in service given by the following system of equations:

$$TD_{US}(t) - TD_{US}(t-1) = \lambda\,(TD^*_{US}(t) - TD_{US}(t-1)) \quad (2)$$

$$TD^*_{US} = h(GNP_{US}, GNP_A, BT, P_{US}, P_A, X) \quad (3)$$

$$TD_{US}(t) = \lambda h(o) + (1-\lambda)TD_{US}(t-1) \quad (4)$$

$$S_{US}(t) = TD_{US}(t)/(TD_A + TD_{US}) \quad (5)$$

where:

- TD_{US} is the demand for service originating in the United States,
- TD^* is the optimal level of demand originating in the United States,
- P is prices,
- X is other demand variables including government policy indicators,
- S_{US} is the share of traffic between country A and the United States.

Equations (1)-(5) describe the demand for telecommunications services between the United States and another country. The resulting partial adjustment model is warranted because analysis of traffic indicates a growth model which depends on the level of demand in the previous period. Lagged adjustment follows from the aggregate nature of the data and the evolutionary process of growth.

There are several potential benefits to be derived from successfully implementing this option. Communications administrations in both the developed countries and the less developed countries can benefit from a reduction in accounting rates that is combined with a modification in the division of accounting rates. In the short run, administrations in less developed countries can anticipate an increase in their settlement payments. They will receive a larger share of settlement payments as a result of the change in the accounting rate division. Moreover, the net traffic flow to less developed countries is increasing so communications administrations in these countries will receive larger settlement payments. At the same time, the reduction in accounting rate levels may bring about lower tariff rates in developed countries and this could stimulate additional demand for service. Thus, administrations in less developed countries will be in a better position to cover the costs of international service and to finance investments in capital equipment. The amount of the increase in payments depends, of course, on the terms of the settlement agreements that are negotiated between administrations and the responsiveness of telecommunications users to changes that have an impact on the demand for service in both countries.

APPENDIX C

by Michael Goetz

A cost study methodology needs to be forward looking in nature in order to capture the actual resource costs associated with economic decisions. Only by asking about the best, prospective use of resources can correct economic decisions be made. The emphasis on forward looking costs insures the best alternative use of labor and capital currently employed even though the costs are estimates.

The costs need to be a comprehensive measure of all the costs of providing international telecommunications service. The cost estimate should include the total expenditure on resources that are used to provide service. Later the costs may be divided in different ways for pricing purposes but they should all be included in the estimate of expenditures. The distinction between long run and short run is not important during the initial stages of cost estimation. Indeed it may be misleading.

A useful way to conceptualize the comprehensiveness of costs is shown in Figure C-1 in the form of a matrix. Total resource expenditure is the sum of elements (a)-(d). This framework is discussed in detail elsewhere.[9]

Figure C-1

	Direct Cost	Indirect Cost
Variable Cost	(a)	(b)
Fixed Cost	(c)	(d)

Costs should not be allocated arbitrarily in a cost study. Cost allocation is necessary in telecommunications because the technology is characterized by joint and common costs. Economies of scope are present. The cost study should assign direct costs to individual services and indicate the costing and pricing consequences of cost allocation decisions. Then issues of cross subsidization and predation can be addressed directly.

A cost study methodology needs to be flexible. Key variables should be defined clearly so that others can assess the analysis and replicate it. Moreover, the methodology should be governed by a unifying principle such as least cost production so the study can be

[9] See Implementation and Scope of the Uniform Settlements Policy for Parallel International Communications Routes, FCC Docket No. 85-204, 51 Federal Register 4736, February 17, 1986.

updated and there is no need to begin anew. Technology changes but the cost study methodology should be sufficiently flexible to allow revision and updating.

The data requirements for an economic cost study are determined by the questions that are asked and the issues that are addressed. In the context of international settlements and the flow of funds between communications administrations in different countries, the economic cost study must be designed to answer three questions. First, what are the direct costs of service? Second, what are the cost implications of producing multiple outputs? Third, what are the interrelationships among cost of service and overall revenue analysis between countries and among countries?

Data on costs, outputs, and input usage in telecommunications technologies are necessary to answer the three questions. The more detailed and richer the data, the more refined the cost study, the more comprehensive the questions, and the more reliable the answers. Even a minimal set of data, however, can be used to provide suitable answers to the basic questions that are being asked. Figure C-2 identifies the basic set of information required to conduct a cost study that provides answers to the questions.

Figure C-2

Information Required to Conduct Cost Study

Definition of cost measures of output, defined cost causatively

Input prices	(wage rate)	(cost of funds)
Input quantities	(hours)	(capital stock)

Measures of output should be defined in a cost causative way so that the outputs are specified in a manner consistent with the use of resources to provide service. For example, do costs depend on call duration or do they depend on the number of calls? If costs depend on call duration, then minutes of use should be the measure of output. If costs depend on the number of calls, then messages should be used.

Input prices and input quantities are critical information to estimating the cost implications of a change in international settlement practices. In the absence of reliable information, a measure of total resource expenditure can be used in place of detailed data on labor and capital.

The quantity and the quality of data that are available are important determinants of the cost study that can be conducted. The data are an empirical link to the world being modelled and the better the data, the more robust will be the results. Moreover, if a statistical study is envisioned, a minimum number of observations is required.

If the minimum requirements for a data set are available, answers to the three basic questions are possible. Direct costs of the service can be isolated. The costs advantages or disadvantages of the production of multiple outputs can be determined either statistically, or as an alternative, by a suitably specified engineering study. Aggregate scale and scope economies can be determined.

If more detailed and richer data sets are available, more detailed issues can be analyzed and more detailed answers can be obtained. Product specific scale and scope economies can be estimated in addition to aggregate scale and scope economies. Incremental costs can be disaggregated into labor and capital components. The sensitivity of labor and capital usage to changes in their prices can be estimated. Measures of productivity improvement may be possible if sufficient time series data are available.

In summary, the economic cost study methodology allows a wide range of policy issues to be addressed even with a modest set of data. A cost causation framework can be used to examine current costing practices and to simulate the interaction of changes in different variables. While the approach is flexible, the availability of data may limit the detail of the analysis. This is the principal shortcoming and it may limit the detail of the answers to the questions.

References

Acton, J.P. and Vogelsang, I. (1988), "Telephone Demand over the Atlantic: Evidence from Country-Pair Data", June 15 (mimeo).

Alleman, J.H., and Stanley, Kenneth E.(eds.) (1990 forthcoming), *Proceedings of the First International Telecommunications Futures Symposium*, Omaha, NE.

Applebe, T.W., Larson, A.C., Lehman, D.E. (eds.) (1990 forthcoming), *Unilateral Pricing of Telecommunications Traffic*, Proceedings of Forum 87

Craver, R.F. (1975), "International Calling at Bargain Rates", *Telephony*, December 15

Craver, R.F. (1976), "A Look into a Crystal Ball at the Future of International Calling", *Telephony*, August 2

Craver, R.F. (1976), "An Estimate of the Price Elasticity of Demand for International Telecommunications", *Telecommunications Journal*, November

FCC (1986), Docket No. 85-204, *Implementation and Scope of the Uniform Settlements Policy for Parallel International Communications Routes*, 51 Federal Register 4736, February 17

FCC (1987), Docket No. 86-494, *2 FCC Record 1022*, January 30

FCC (1988), *Report and Order and Supplemental Notice of Inquiry*, March 25

Goetz, M.L. (1989), "Cost Allocation Techniques and Pricing Alternatives: Crossing the Great Divide", in Alleman, J.H. and R.M. Emmerson (eds.), *Perspective on the Telephone Industry: The Challenge for the Future*, Ballinger Publishing Company, Cambridge

Khadem, R. (1976), "An Econometric Forecasting Model of the Demand for International Voice Telecommunications from Canada", *Teleglobe Canada*, October

Kumins, L. (1967), "An Estimate of Price Elasticity of Demand for International Long-Distance Telephony", *Telecommunications Journal*, June 1

Lago, A. M. (1970),"Demand Forecasting Models of International Telecommunications and Their Policy Implications", *Journal of Industrial Economics*, November

Rappoport, P.N. (1977), "Forecasting the Demand for International Message Telephone Service", March

Rappoport, P.N. (1979), "A Forecasting Model of MTS Demand between the United States and the United Kingdom", *Federal Communications Commission*, February

Stanley, K.B. (1988), "The Balance of Payments in International Telecommunications Services", *Federal Communications Commission*, December 12

Stanley, K.B. (1990 forthcoming), "Trends in the International Telecommunications Industry, 1975-1988", *Federal Communications Commission*

Taylor, L.D. (1980), *Telecommunications Demand: A Survey and Critique*, Ballinger Publishing Co., Cambridge.

Willig, R.D. (1976), "Consumer's Surplus Without Apology", *American Economic Review*, vol. 66;589-597.

Yatrakis, P.G.(1972), "Determinants of the Demand for International Telecommunications," *Telecommunications Journal*, December

Comment Chapters 5 and 6

Hans-Peter Gassmann

OECD
Paris

Both papers are very interesting, as they shed some light on so far not too well explored aspects of telecommunications. They have two aspects in common: firstly, they are both forward looking and pioneering, they explore certain areas were so far we haven't had any kind of more detailed analysis and secondly both authors complain about the absence of hard, internationally comparable data - not surprisingly, as this is a general feature of research on so far little known fields. Their subjects, however, are rather unrelated; the first makes some preliminary comparisons of R&D expenditures in telecommunications among developed countries, whereas the other is concerned with the telecommunications settlements between developed countries and developing countries. The first is very empirical, whereas the second is more on methods and procedures. For these reasons, I will discuss them separately.

R&D in Telecommunications

Dr. Schnörings's paper develops a typology of relations in telecommunications R&D between telecommunications carriers, public and private ones, manufactureres of telecommunications equipment, and research institutes and universities, and draws some conclusions on the relative strength and weaknesses for future international competitiveness. He points out that the telecommunication sector is very R&D intensive in the most of the countries compared, which earmark between 10 - 13 % of their total national R&D expenditure for this sector. This is a remarkable amount, which points to the great stategic importance of this sector both as an infrastructure and also, at least for the United States, the UK, and France, as a major element of national security. He shows that for all the countries examined, government has a great influence on the directions of R&D in this field.

Another factor to be noted is that the contributions of Universities to research in telecommunications, as measured by the share of research publications, is remarkably similar for the UK (32 %), Germany (34 %), and the United States (29 %), whereas both in France and Japan the Universities seem to have a much lower role.

Mr. Schnöring's observation that it would appear that divestiture in the United States has, at least for the period observed, substantially increased the total R&D effort in telecommunications in that country invalidated some of the fears which were widely expressed before 1984, namely that divestiture would actually weaken the R&D performance in telecommunications of the United States. Two questions could be raised here: the first is whether this is still true for the more recent period (after 1987), and second, whether there has been an adverse impact on fundamental research in favour of more applied research, as US companies are now in a more competitive situation, and view applied R&D as a competitive element.

Another point worth to be emphasized is Mr. Schnöring's description of the negative effect of too great symbiotic relations between the dominant carrier and national manufacturers, in the absence of competitive pressure, in the case of the development of EWS in Germany and System X in the UK.

Finally, in his conclusion Mr. Schnöring points to the strong influence of the French and Japanese governments on the level and the direction of R&D in telecommunications, and that this tends to create market distortions. Here it should be pointed out that this situation is not particular to the telecommunications sector, as this can be observed in other fields as well.

This situation can, at least partially, be explained by some deeper, almost cultural reasons: in France, there has been a tradition that the government embarks in what is called "grands projects". This is a strong tradition which may be explained by the fact that it is a very centralized system. In Japan there is a long tradition of "human networking", which manifests itself in particularly close ties between government, industry, and research institutions, and consensual prority setting. The result of these different traditions is the same: in both countries a longer-term view is taken, less influenced than in other countries by short-term, more commercial interests.

International Settlements in Telecommunications

Pr. Alleman's paper deals with the growing deficit in the United States in the telecommunications services balance of payments, and the demand of some developing countries to change the classical 50-50 % division of the proceeds of telecommunications service revenue between two countries. He demonstrates that the US telecommunications services deficit was 1.7 billion US-$ in 1987, and was growing by 22 % per annum. He terms this deficit as "significant". The question must be raised how significant this figure is; after all, for a country as large and as developed as the US 1.7 billion might not be very large; it might be useful to plot this figure against the total balance of payment figure, in order to see how both series develop over time. My estimation is that the telecommunications services deficit may be about 1-1.5 % of the total deficit.

Mr. Alleman does not offer explanations why this deficit is growing. So another question is: why does it grow? I want to offer two hypotheses, which might be worth to be explored further. First, it has been mentioned elsewhere that the main effect of divestiture in the US was in rate rebalancing, and that over the last 4 years the prices of long distance calls have been declining by 30 %. This has probably been true also for international calls; if so, the price differential between the US and other countries, where such a reduction has not taken place, has grown, and this must have had an influence on the deficit, as the incentive to originate calls from the US has increased. Second, because of this growing difference in prices, a dramatic increase of reverse-charging calls from other countries to the United States has probably occurred, as well as an increase of very short calls of the type "call me back - will reimburse you later."

It should be noted, however, that the years 1986 and 1987, for which he has observed the increase in the deficit, coincide with the start of mass utilisation of international analogue facsimile transmission. Here reverse charging or short calls for call-back are not possible; a fax transmission from, say, Brazil to the US has to be paid by Brazil. This development may actually have had a dampening effect of the deficit situation.

Concerning the request by less developed countries to change the division of the revenue split from 50-50 to for example 40-60, Mr. Alleman mentions as a reason the high costs of international calls on their side. There are no hard data to substantiate this assertion;

one may speculate that partly this might be due to the high cost of imported equipment, the relatively higher costs of maintencance (spare parts, lack of training, need to call on foreign specialists, etc). On the other hand, the much lower wages even of specialists of these countries might partly offset these higher costs of equipment; so it is an open question whether the less developed countries' claim is correct.

Finally, Mr. Alleman proposes that a lowering of accounting rates could be negotiated in return for a different revenue split; this proposal has great merit. But the question must be asked whether an annual drop by 10-20 % of the accounting rate which he proposes is realistic. In my view, this percentage is far too high, and a figure of say 5 % might be appropriate.

Conclusion

Despite different subjects, both papers relate to the distorting effects of expenditures, R&D expenditures and settlement arrangements, on international trade in goods and services. As telecommunications have a strategic importance for international trade, both as revenue-generating in their own right as well as facilitators for other traded goods and services, it is important to continue generating more international comparisons and research on these topics, which eventually will influence policy-making. The authors of these papers need to be commended on having started shedding more light on these subjects.

7

Multinational Banking, the Theory of the Division of Labor and the Information and Communications Technologies

Bernhard Wieland[1]

Faculty of Social Science
University of Cologne,
Cologne

1. Introduction

Telecommunications economics as an academic field developed simultaneously with the regulatory battles in the last two decades. No wonder therefore that up to now it has been mainly preoccupied with questions like natural monopoly, tariffs, predatory pricing, rate of return regulation or standardization. Apart from employment effects, the more general effects of the new information- and communication technologies (telematics, for short) on society have up to now largely been left to political scientists and sociologists. Only a minority of telecommunications economists has up to now directed its attention to the effects of telematics on the structure of firms, industries or the economy at large. With the regulatory battles largely over (to venture this optimistic hypothesis) it is to be hoped

---

1 I wish to state my intellectual indebtedness to Professor C.C. von Weizsäcker. In various publications and in private conversations he has stressed the importance of the theory of the division of labor and pointed out its relation to the information and communication technologies (see e.g. his 1988). Needless to say that he bears no responsibility for whatever shortcomings this paper may have.

that more economists will turn to the study of the wider impact of telematics on the various national economies and the world economy.

This paper attempts to contribute to this endeavour in one special field: financial markets. The study of these markets is of interest because information processing and information transmission are of special relevance in finance. The hypothesis has been ventured that the financial sector is a "vanguard industry" in the use of telecommunications (Barras, 1988). Financial markets, however, are largely driven by the activities of the main market makers, banks. The paper therefore deals exclusively with banks.

The main thesis of the paper is that the deeper cause for the multinationalization of banking which has occurred since the early 60s is the cost reducing effect of telecommunications. Of course it has always been acknowledged that telematics has played an important role in the internationalisation of financial markets. But usually telematics is considered to have been just one factor among others. This paper argues that without telematics the internationalisation of financial markets could not have occurred. The information and communication technologies provided the necessary transaction infrastructure which formed (and continues to form) the basis of this process. Other hypotheses of the internationalisation of financial markets are not satisfactory as long as they do not incorporate this element.

In its second part the paper turns to the classic question of the division of labor. Here the expression "division of labor" is meant to refer not only to the division of tasks within firms. It is meant to refer also to the division of activities between firms in a given economy or even in the world economy. Economic theory predicts that growing markets will be accompanied by a growing division of labor. The multinationalization of banking sheds some interesting light on this proposition. We contend that banks will grow more and more into internationally integrated entities but will increasingly become more specialised. Thus the international division of labor will take the form of more specialisation of worldwide operating but highly specialized banks. This is in contrast to the currently fashionable idea of "financial supermarkets" ("Allfinanz").

2. Technology and the Extent of the Market

In 1776 Adam Smith identified the division of labor as a major source for the "wealth of nations". However, "the division of labor is limited by the extent of the market". "When the market is very small, no person can have any encouragement to dedicate himself entirely to one employment, for want of the power to exchange all the surplus part of the produce of his own labour, which is over and above his own consumption, for such part of the produce of other men's labour as he has occasion for. ..... In the lone houses and very small villages which are scattered about in so desert a country as the Highlands of Scotland, every farmer must be a butcher, baker, and brewer for his own family." (Smith, 1776, p.13)

But the extent of the market, in turn, is determined by the available technologies of transportation. "As, by means of water-carriage, a more extensive market is open to every sort of industry than what land-carriage alone can afford it, so it is upon the sea-coast, and along the banks of navigable rivers, that industry of every kind naturally begins to subdivide and improve itself, and it is frequently not till a long time after that those improvements extend themselves to the inland parts of the country. A broad wheeled waggon, attended by two men, and drawn by eight horses, in about six weeks time carries and brings back between London and Edinburgh near four ton weight of goods. In about the same time a ship navigated by six or eight men, and sailing between the ports of London and Leith, frequently carries and brings back two hundred ton weight of goods." (Smith, p.14) Due to its higher productivity water-carriage therefore is much cheaper than land-carriage. And the cheapness of transport determines how much trade there is between two places. "Were there no other communication between those two places, therefore, but by land-carriage, as no goods could be transported from the one to the other, except such whose prices was very considerable in proportion to their weight, they could carry on but a small part of that commerce which at present subsists between them. ... There could be little or no commerce of any kind between the distant parts of the world. What goods could bear the expense of land-carriage between London and Calcutta? ... Those two cities, however, at present carry on a very considerable commerce with each other, and by mutually affording a market, give a good deal of encouragement to each other's industry." (Smith, p.14 and 15)

The societies of our own day are frequently described as "information societies". Be that as it may, it can hardly be doubted, that the transportation of information (in particular of processed information) has become of great importance for our national economies and the world economy. The last decades have seen the introduction of the new information- and communications technologies, which have dramatically lowered the costs of processing and transporting information between two places. This paper offers the conjecture that Smith's theory is applicable here too. I venture the hypothesis that the internationalization of markets which we have seen in the last 20 to 30 years is the result of telematics. To be more precise: I contend that telematics was a necessary condition for the internationalization of markets or that without telematics this process would not have occurred.

The internationalization of banking illustrates this conjecture very well. Suddenly in the early 60s banks began to "go international" in numbers that had not be seen before. This does not presume that up to the 60s there had been no international banking at all. Of course international banking goes back as far as the early renaissance (Kindleberger, 1984). The new quality was that suddenly almost *all* banks began to take on an international perspective, not only a select few like the Medici- or the Fugger-bank in earlier times. Today it seems as if participation in the international business has become a *conditio sine qua non* for modern banks.

Several explanations have been offered for this phenomenon. I shall discuss the most important of them below. All these theories, however, either overlook the effects of telematics or consider them to be of secondary importance. As opposed to this I propose to stay with Smith and to consider transportation technology (this time for the transport of information) to be the main determining factor for "the extent of the market".

One frequently encounters the notion that telecommunications can not be so important after all, because its share of total costs of an enterprise is low. (Estimates range from 2 to 10% in banking). This view is based on a misunderstanding. Telecommunications has a large influence on *other* cost blocks which may have a much larger share in a company's budget. Take as just one example the opportunity costs of time of high paid managers or technical staff. These individuals spend a lot of their time travelling between firm locations instead of sitting in their offices and performing their regular functions. This causes high opportunity costs for firms. Companies report that the introduction of video conferencing has lowered these costs considerably. Companies also state that these opportunity costs of time are far more important than pure travelling costs.

Therefore what is involved here are not only reduced telecommunications costs in the form of falling tariffs and falling equipment costs. Far more important are the secondary effects of telematics which usually are described as reduced transaction costs. The most important effect of telematics has been to reduce the costs of transaction over long distances. And this, I contend, has induced companies to switch from a national to an international business perspective.

It is another question whether it is possible to obtain reliable data on the cost reductions caused by telematics in that way. At least my own efforts have been not very successful so far. Nevertheless it is possible to offer some sketchy evidence on this point as I shall show in section 4. It cannot be denied, however, that up to now there is no systematic empirical research available to shed light on the transaction cost reducing effects of telematics. This is an open field of research.

3. Internationalization and Multinationalization of Financial Markets

During the last two decades international banking has shown unprecedented growth rates. Big banks today consider the establishment of a worldwide network of branches or subsidiaries to be of high strategic importance. But also smaller banks try to participate in this development by establishing presence at least in some of the most important financial centers. The larger banks have built up large worldwide telecommunications networks to link their domestic and international branches and subsidiaries.

From the perspective of economic theory two questions arise. First: Why have financial markets suddenly become "global"? Second: Why have banks expanded their business via direct investment (that means the establishment of own branches or subsidiaries) and not via international market transactions with foreign correspondent banks? Until the late fifties and early sixties doing business with correspondent banks, not direct foreign investment, was the rule.

These two questions are not identical. The first relates exclusively to the fact that since the 60s banks operate more and more with an international perspective. The second relates in addition to the *way* in which this internationalisation has occurred. In the following I shall refer to the first question as the "internationalization question" and to the second as the "multinationalization question".

Let us first offer some empirical evidence concerning the internationalization of financial markets. A first impression of the growth of international financial intermediation is provided by the following figure 1. It shows three variables: world gross product, world exports of goods and services, and the volume of international banking. (The last variable is represented by the volume of the Euro-Markets measured according to the Morgan Guarantee concept.) As can be seen international banking exhibits substantially stronger growth than the other two variables. Only since the beginning of the 80s has the growth somewhat levelled off due to the recession and the world debt crisis.

Figure 1: International Banking, World Exports of Goods and Services, and World Output, 1964-85

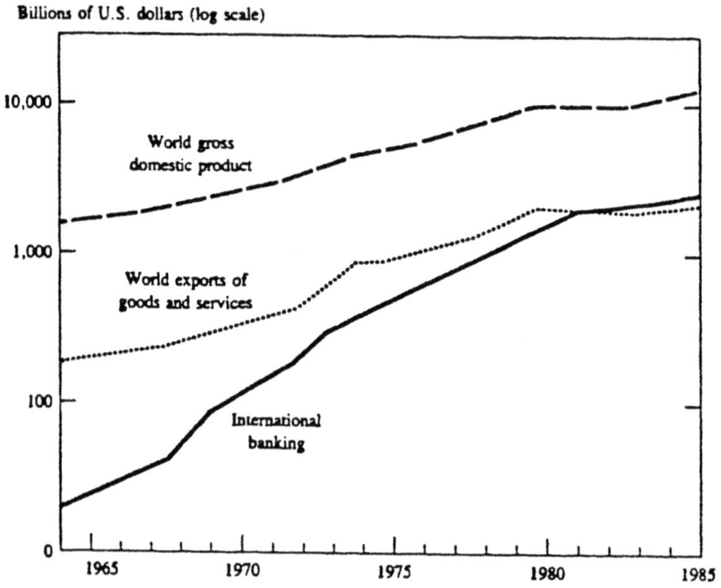

Source: Bryant (1987)

We can now look at the Euro-markets in isolation. The following figure 2 shows the development of the Euro-credit-market and the Euro-money-market between 1966 and 1983. The variables shown are business volumes according to differing measurement concepts. It is not necessary here to go into the details of the measurement of these magnitudes. Roughly the lower part of the figure refers to the European segment of the Euro-credit- and the Euro-money market and the upper part of the figure to the so called extended reporting area of the Bank for International Settlements which includes the overseas Euro-centers as well. Moreover, for each variable net- and gross volumes are shown. The net-variables are obtained by subtracting the interbank transactions from the gross variables. As can be seen, all varaiables exhibit substantial growth rates which lie between 25% and 30% annually.

Figure 2: Euro Credit-Market and Euro Money-Market 1966-1983

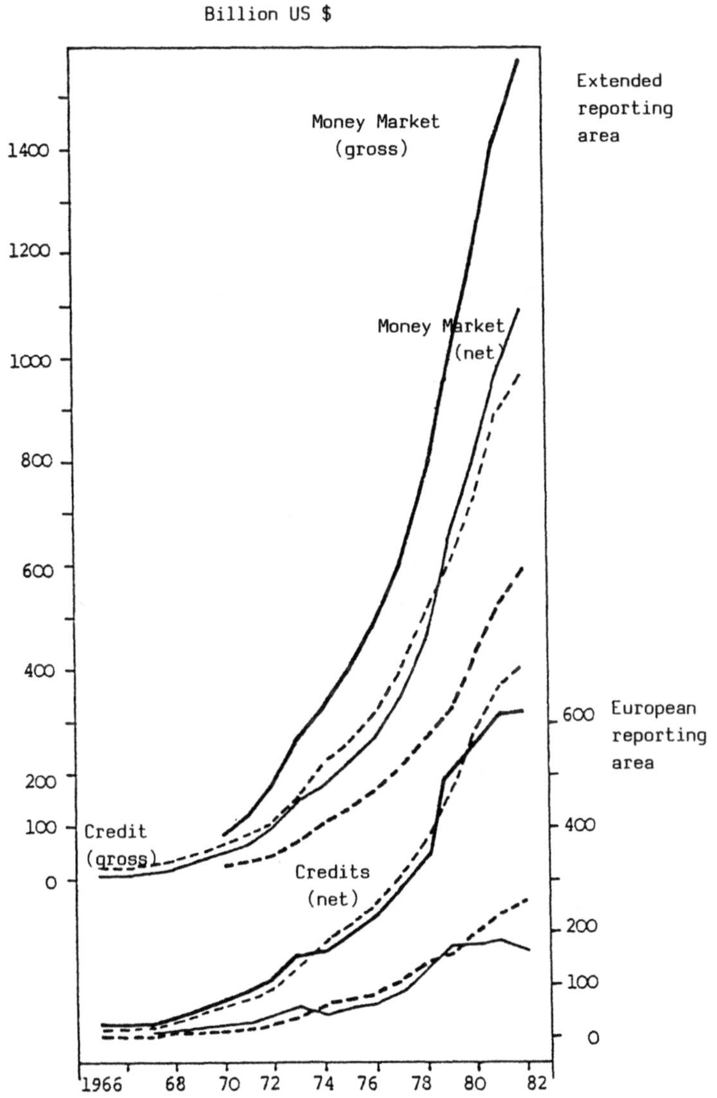

Source: Dennig (1987)

Of course there are still other indicators of the growing internationalization of financial markets. From the perspective of economic theory one of the best indicators is the degree to which the national financial markets reflect the law of one price. There are two kinds of empirical studies which seem to confirm this law for international financial markets. The first looks into the prices for international securities. If adjustments are made for issuer, maturity, rating etc. it seems that the yields of such securities are similar in the Euro-market and the various national markets. This suggests a sufficient amount of arbitrage activities in international markets which wipe out yield differentials. The second kind of empirical studies tests covered interest rate parity conditions and the existence of covered interest arbitrage profits. Again there seems to be overwhelmimg evidence that international markets are now fairly integrated (Levich, 1987, p.41).

Let us now turn to some empirical evidence concerning the multinationalization of banking. In the fifties, when transnational banking activities consisted mainly in the financing of international trade national banks mainly effected this business via foreign correspondent banks. Subsequently, as international business expanded and as the Euro-markets and off-shore centers developped in the 60s and 70s the transactions with correspondent institutes gave way to closer cooperations, to take-overs or, finally, to the establishment of own branches or subsidiaries. The following table 1 gives an indication of this development. It shows financial institutions or banking offices owned by foreign banks in selected host countries in the time from 1960 to 1985. Again there are substantial growth rates involved. If branches or offices in the off-shore centers were included the groth rates would be even greater.

Table 1: Institutions or Offices Owned by Foreign Banks Operating in Selected Host Countries, December 1960-June 1985

Host country	December 1960	December 1970	December 1980	June 1985
		Number of institutions[a]		
Belgium	14[b]	26	51	57
Canada	0	0	0	57
Italy[c]	1	4	26	36
Netherlands	n.a.	23	39	40
Switzerland	8	97	99	119
United Kingdom	51[d]	95	214	293[e]
		Number of banking offices[f]		
France	33	58	122	147
Germany	24	77	213	287[g]
Japan[h]	34	38	85	112
Luxembourg	3	23	96	106
United States	n.a.	n.a.[i]	579	783[j]

Source: Bank for International Settlements, Recent Innovations in International Banking, report prepared by a study group established by the central banks of the Group of Ten countries (Basel: BIS, 1986), p. 151.

n.a. Not available.

a. Figures are for the number of foreign-owned banking institutions ("families") operating in the host country through branches or majority-owned subsidiaries unless otherwise specified. An institution maintaining offices at several locations is counted only once. Figures are for end of month.
b. December 1958.
c. Figures are for branches only. At end of June 1985, five foreign-owned subsidiaries were operating in Italy.
d. December 1962.
e. The figure for June 1985 is 357 if joint ventures and consortium banks are included.
f. Foreign-owned banking institutions represented by more than one entity in a host country are counted more than once. Figures are for end of month.
g. At end of June 1985 these offices represented 95 different foreign-owned banking institutions.
h. Figures are for branches only. At end of June 1985, 76 different foreign-owned banking institutions were operating in Japan.
i. In the early 1970s about 50 offices of foreign-owned banking institutions were operating in the United States.
j. At end of June 1985 these offices represented approximately 350 different foreign-owned banking istitutions.

Source: Table reproduced according to Bryant (1987)

4. A Case Study[2]

To illustrate the role of telecommunications in the process of the multinationalization of banks I very briefly sketch the case of a large German bank. This bank is heavily involved in international banking. It has five foreign subsidiaries located in Luxemburg, the Netherlands, Switzerland, Singapore and the United States. Furthermore, it has 13 branch offices and 17 representative offices in Argentina, Australia, Bahrein, Belgium, Brazil, Canada, China, Cayman Islands, Denmark, Egypt, Finland, France, Hongkong, Indonesia, Iran, Italy, Japan, Mexico, Spain, Southern Africa, Turkey, the USSR, the United Kingdom, the United States and Venezuela. Moreover it holds shares in 30 other financial institutions and other companies abroad.

In order to coordinate all these foreign activities (and of course its domestic activities) the bank has installed an elaborate telecommunications network (see figures 3 and 4 below).

This network was set up in 1983 and consists of about 30,000 kilometers of leased lines (HfD) in Germany and 27,000 kilometers abroad. In conformity with the bank's organizational structure the network exhibited at first a star-topology with the center located in Frankfurt. This topology, however, created reliability and security problems. Every communication between two points had to pass through the center. There was therefore the danger that in case of a break down of data-processing at the center all business would have to stop. Experts today predict bankruptcy within a few days if such a breakdown lasts for an extended period. Therefore the bank decided to establish a second back-up concentration point at Düsseldorf.

The network connects the bank to all important financial centers of the world. The link from Frankfurt to New York is a 64KB satellite link. New York is the concentration point of the US-network. It provides access to the network nodes at Chicago, Atlanta and Los Angeles. The connection to the Far East is established by means of a direct route to Tokio and a second one to Hongkong (both by satellite) with a network node located at Singapore which is linked to Hongkong by seacable. The routes to the US are served by an US-carrier (TRT), and the Far East routes by KDD (the links to Tokyo and Hongkong), Cable and Wireless and Telecom Singapore (the link from Hongkong to

2 This chapter is a radically shortened version of a case study which the author and G.Pfeiffer have performed for a joint project of the OECD and the Berkley Roundtable on the International Economy (BRIE) at the University of Berkley. The German part of the study (which contains three other case studies) is soon to appear in book form.

Figure 3: National telecommunications Network of Commerzbank

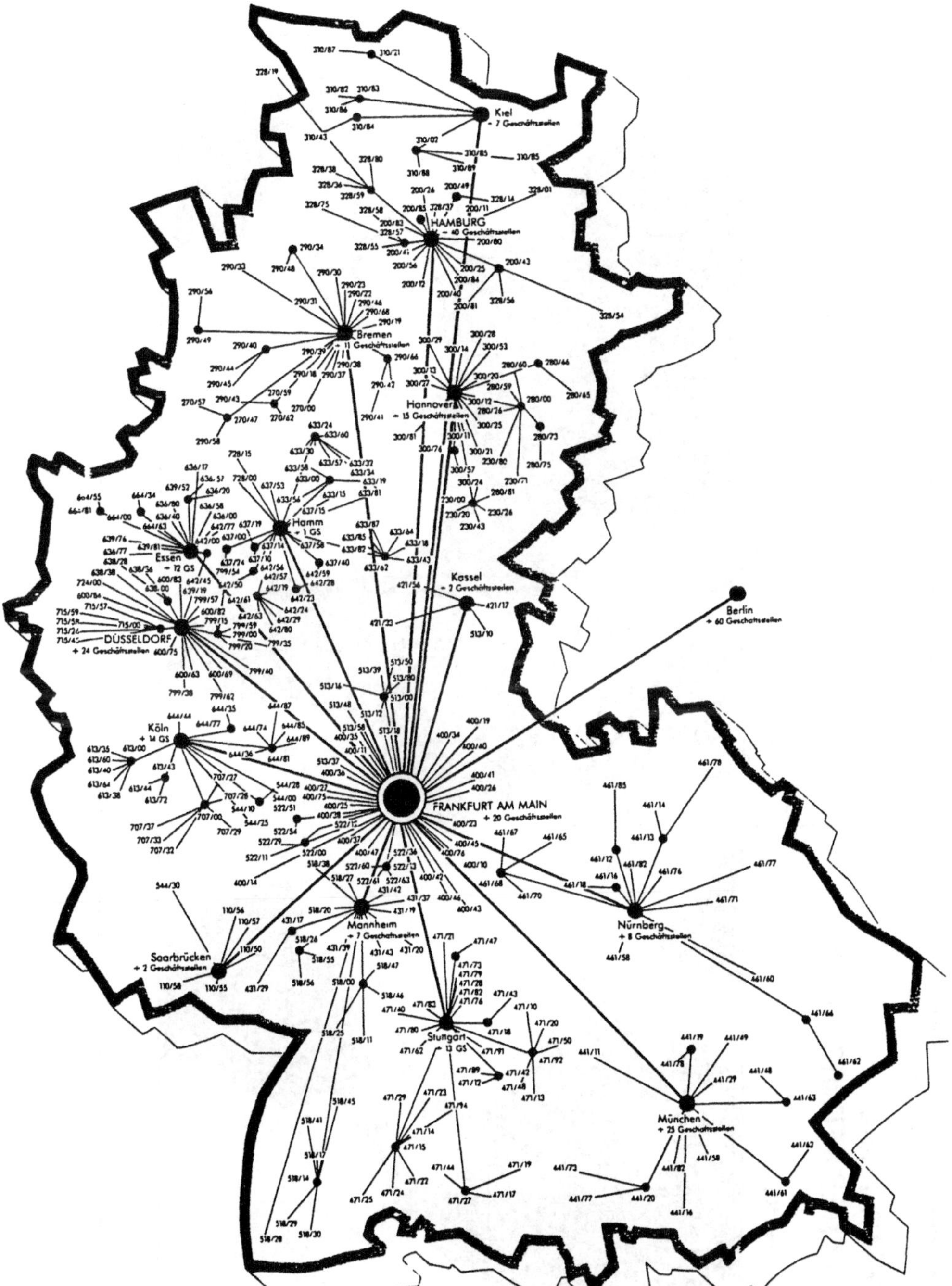

Figure 4: International Computer Network of Commerzbank

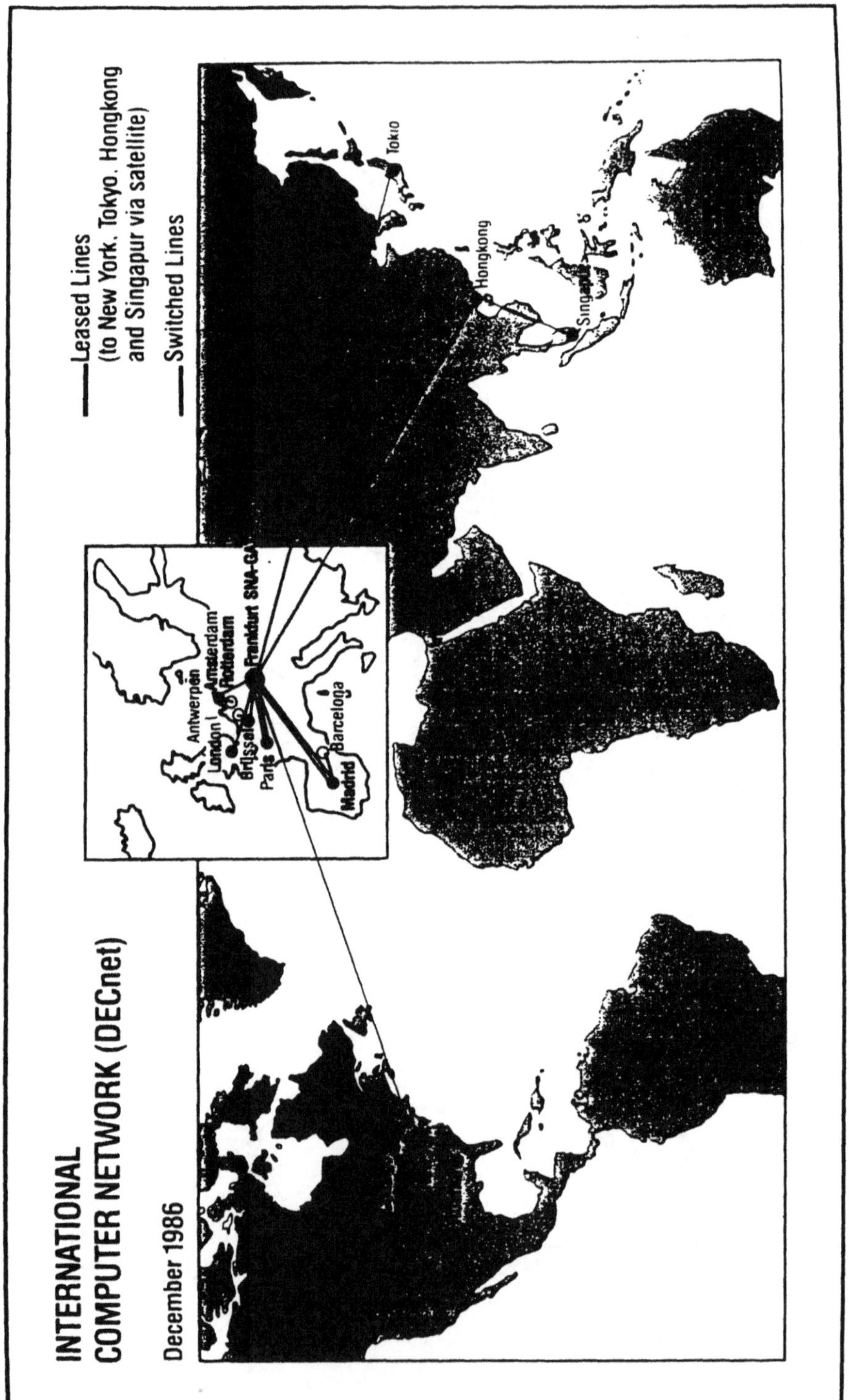

Singapore). Moreover the bank is, of course, connected to the SWIFT-network, for which it spends 25 million DM annually.

All in all the network connects more than 10,000 terminals and their peripheral devices (more than 5,000). An important application of the bank's worldwide network is an electronic mail system over which around 140,000 memos are exchanged daily. The bank runs about 5,000 applications programs on its data processing machines of which 2,500 can be used in conversational mode.

A network of branches and affiliates at strategic locations all over the world, connected via a telecommunications network like the one described, offers a bank an enormous competitive advantage. Countries all over the world differ in the regulations they impose on the financial sector. In particular, the so called off-shore centers are known for their low degree of regulation and their low taxes. Thus, with a large network of branches and affiliates and a corresponding telecommunications network banks can shift activities to those locations which offer most advantages. Moreover, with such a network funds can quickly be transferred from one branch to another according to profit opportunities and temporary shortages or surplusses. The same holds true for financial information. Information that has been generated somewhere in the network (e.g. about an important multinational customer) can easily be reused at any other place in the network. Thereby economies of scale in collecting and using information can be exploited. Moreover, these networks allow ready access to all national capital markets. They also allow participation in 24 hours trading. The responsibility for the bank's portfolio management can be shifted around the globe from one branch to the next from time zone to time zone if the bank has operating units all over the world.

These are only the most important advantages. Nevertheless, it should be already plain from this enumeration that telecommunications today is of overwhelming importance for international banks. A bank which uses telecommunications in these ways has tremendous cost advantages over its competitors. As mentioned already these cost savings are not reflected in the cost share which telecommunications has in the total costs of a bank.

5. Theories of Internationalization

Several explanations have been offered for the internationalisation of financial markets. These explanations can roughly be grouped into three categories: theories which emphasize regulatory factors, theories which see banks as special multinational enterprises, and eclectical theories. I comment on every strand of theory very briefly.

The first kind of theory attributes the growth of financial markets to regulatory factors. Consider for instance the so called Euro-Market. This market derives its importance to a very large degree from the fact that the usual national banking and monetary regulations do not apply to "Euro-business". This is achieved by simply conducting a financial transaction denominated in a certain currency outside the country of that currency. Thus, a dollar transaction conducted via a London bank is a Euro-Dollar transaction. Likewise a transaction in German marks taking place in Luxemburg is a Euro-DM-transaction. By this artifice the transactors operate (so to speak) in a regulatory no-man's-land. This means that costly regulations like for instance minimum reserve requirements can be avoided. The same holds true for banking in so-called "off-shore" centers (like the Bahamas or the Cayman Islands) which, in addition, frequently offer considerable tax advantages.

It must be asked, however, whether the desire to escape regulations is really enough of an explanation for the internationalisation of financial markets - or rather whether the explanation goes deep enough. First, in many cases it was possible to evade regulations *at home*, for instance with the help of "financial innovations" like monetary market funds (in the case of interest ceilings)[3]. Second, the opportunities for the creation of an Euro-market and the creation of off-shore banking places were there before the early 60s, at least as far as the dollar segment of the market is concerned. In fact, US banks started operating in London already shortly after World War II. (For the other currencies this statement needs certain qualifications. Restrictions on capital movements continued to apply in some currencies for quite a while after World War II. Still, it is doubtful whether freedom of capital movements without telematics would have led to the internationalization of financial markets.)

So much for the argument that the internationalisation of financial markets must be attributed exclusively to regulatory factors.

The next group of hypotheses is closely connected to the theory of the multinational enterprise (MNE)[4]. Banks are seen as a special form of MNE exploiting, for instance, "special advantages" in the sense of Hymer (1960/1976). Such a special advantage may be, for example, specific knowledge about a certain customer who has decided to take

3 These innovations, by the way, also depend heavily on telematics (Berlin/Philips, 1985).
4 For an excellent short survey see Teece (1985).

up operations abroad. In such a case - so the argument goes - banks have a certain advantage in doing business with such a client due to their long standing business relationship with him. Therefore banks decide to "accompany" their customer abroad. Considering that the customers in question are in most cases large firms the banks also want to make sure that they do not lose these customers to foreign banks. This can only be done by establishing own branches abroad. Other variations of this theory attribute special advantages to the better access of a bank to its home currency or to other factors like process innovations etc.

This kind of theory, however, suffers from certain inconsistencies. In many cases, for instance, banks were rather leaders than followers. More important, however, is the fact that again these theories do not explain why the internationalisation of banking took place in the last two decades and not earlier. In the case of the "follower hypothesis" this objection may be answered by pointing out that before that time there were simply not so many multinational enterprises. But this immediately raises the question why suddenly in the last two decades this has changed and more and more firms have become international. Again I venture the hypothesis that this was largely a result of better communications possibilities and reduced transportation costs.

The third kind of explanation for the internationalisation of banking is offered by an application of the so called "eclectical" theory of the MNE (Dunning, 1977) to banking. As its name already indicates this theory tries to combine elements of the afore mentioned theories into a new theoretical structure. One particular variation of this kind of theory seems to have a direct connection to telecommunications (Kogut,1983). This theory takes its departure from the fact that the large multinational enterprises are establishing worldwide *networks* of branches and subsidiaries. As we have already seen this is particularly true for banks which install networks, which span all continents and all time zones (see also Glüder, 1988). The networking theory uses elements of all the other theories to explain why it is highly advantagous for banks to have such a network. I have already described some of these advantages at the end of the case study above. Again it has to be said, however, that these advantages have become exploitable only because a new production technology has arrived: telematics.

6. The Economic Theory of the Division of Labor

We have already mentioned Adam Smith's famous theorem that "the division of labor is limited by the extent of the market". In a famous paper written in 1951 George Stigler reexamined Smith's theory and applied it to the division of labor between firms. He started by observing that usually firms embrace many activities which in principle could be spun off and left to the market, in particular when these processes exhibit economies of scale. It may be, however, that the sales of a firm's product are too small to support new firms specializing in these activities. The demand for a certain intermediate product in the automobile industry, for example, may be too small to support a specialized fabricator. Therefore as long as the market for automobiles is small an automobile firm has no choice but to manufacture the intermediate product itself. But as soon as the market for a certain industry grows the potential for spinning off activities grows as well. With expansion of the market for the firm's end-product the magnitude of such activities may become sufficient to permit a firm to specialize in performing it. The firms will then spin-off the process into a seperate enterprise. This will happen in particular when the activity in question exhibits diseconomies of scope with the other activities of the firm. To quote Stigler himself: "Actually, many processes will be rival: the greater the rate of output of one process, the higher the cost of a given rate of output of the other process or processes. Sometimes the rivalry will be technological (as in many multiple-product firms), but almost always it will also be managerial: the wider the range of functions the firm undertakes, the greater the tasks of coordination."

Thus Stigler has shown, how market growth leads to increased division of labor between firms. There is, however, one gap in his argument: For the division of labor to increase it is not necessary that activities must be spun-off into autonomous enterprises. Adam Smith's own classical example of the pin-factory makes this strikingly clear. What Smith describes is how division of labor *within* the pin-factory increases productivity. There is no mention of activities being spun-off into autonomous enterprises. It is therefore necessary to be a little more explicit on the spin-off decision of the firm.

If, for instance, there are strong economies of scope between the candidate for seperation and the other activities of the firm and if these economies of scope are not outweighed by managerial diseconomies then a spin-off is unlikely.

Likewise, if the firm thinks that the costs for securing a reliable flow of intermediate processes or products from the spun off unit via market transactions outweigh the costs of coordinating the business unit with the other divisions of the firm it will decide against the spin-off. In the reverse case the spin-off will take place. Thus, all else equal, in this case it is the relation between coordination and transaction costs which ultimately decides whether division of labor will be effected within the firm or via a new independent business.

Finally, the spin off will not take place, if the market which is established by this act (or which exists already) suffers from market imperfection. This may in particular be the case when the exchange of information is involved or when informational asymmetries play an important role. Another important source of market imperfections may be regulations.[5]

Therefore in growing markets industries will exhibit increasing division of labor unless

(i) economies of scope do play a dominant role and generally outweigh diseconomies in management,

(ii) transaction costs are substantially higher than coordination costs,

(iii) there are important market imperfections (like an asymmetric distribution of information, natural monopoly, external effects, regulations or ruinous competition).

It can be shown that these three categories can be reduced to one: the presence of important transaction costs[6]. This was already seen by R.H. Coase in 1937 and has become one of the main topics of "transaction cost economics" (see e.g. Williamson, 1985).

7. Banking and the International Division of Labor

It is puzzling at first sight that the internationalization of banking occurs mainly via direct foreign investment (the establishment of own branch offices or subsidiaries) and not (or at least much less) via transborder transactions with foreign correspondent banks. The puzzle lies in the fact that in the case of financial markets substantial growth of the market (into worldwide dimensions) has *not* been accompanied by more market transactions between increasingly specialized agents. How can this development

5 Of course Stigler himself notes in his article that market imperfections may constitute barriers against spin-offs. He mentions price regulation and monopoly as examples. Also, in a concluding note to a reprint of his article published in 1968 he notes that "reference should have been, and now is, made to R.H. Coase, 'The Nature of the Firm'". This probably has to be interpreted in the way that Stigler would have incorporated transaction costs in his article had he known of Coase's rather neglected contribution already in 1951.

6 Economies of scope, for instance, arise mainly when an activity which exhibits economies of scale can be used for more than one purpose (Schmalensee 1988). Clearly this activity or resource could in principle be acquired via the market. What prevents this from happening are in most cases high transaction costs. Likewise, information asymmetries could in principle be reduced if the less informed partner were prepared to incur costs for independent experts or other ways of reducing his lack of information. That problems of natural monopoly and of external effects can be traced to high transaction cost has become conventional wisdom since the pioneering contribution by Coase (1960). A further elaboration of this point is given in Dahlmann (1979).

reconciled with the theory that has been just described? Moreover, how can it be reconciled with the fashionable opinion that the IT technologies will cause firms to disintegrate?

Asking this question is simply another way of asking oneself why multinational banks exist at all. As we have seen, it may be the case in certain circumstances that market transactions are in many cases too expensive or simply not feasible. This happens, for instance, if the specific good or service that a firm wants to offer in the foreign market involves some special kind of know-how. In such a case direct investment is a better way to appropriate the economic rents connected to this know-how than a market transaction like licensing. Costs of negotiation, control and enforcement (i.e. transaction costs) would be prohibitively high if the firm wanted to make absolutely sure that its partner does not cheat on the arrangement[7].

Banks seem to fit well into this line of reasoning. The basic productive input in banking is information (e.g about a certain customer). It is often said that what banks are really engaged in is the production and sale of information. It is well known since Arrow's famous contribution in 1962 that markets in information do not work well if at all. One reason for this is that once information is disclosed it is virtually completely out of the control of its former owner. In order to appropriate the rents associated with a certain piece of financial information banks are therefore not likely to license a foreign correspondent bank to use this information. The bank will consider the establishment of an own branch in the foreign country to be a safer and cheaper way to appropriate these rents. Moreover, by establishing a worldwide network of foreign branches banks can "internalize" all the advantages which have been described above. For these reasons there will be no division of labor between different nations in banking. Banks are likely to remain large multinationally integrated entities. Progress in telecommunications- and data-processing technology will not change this situation.

But Stigler's theory predicts another kind of division of labor. According to this theory it is likely that banks may not remain "universal banks", which offer all types of banking services under one roof. This applies especially for German banks, which in most cases are universal banks. In other countries there is usually a separation between commercial banks, investment banks and savings and loans institutions. In view of Stigler's theory we should expect this separation to become the rule. But even within these broad categories there will probably be increased specialization, for instance according to

---

7 This is the main thrust of the "internalization" theory of the multinational enterprise. See Teece (1985).

customers or financial instruments. An example is already provided by mutual monetary market funds. These funds offer a special financial product which combines the characteristics of a checking account and an investment funds. The managers of the funds invest in money market titles. Accordingly, the shares of a particular customer in the funds can quickly be exchanged into money. This makes it possible that the customer can also write checks against his shares in the fund[8]. It is probable that we shall see more specialized institutions like these funds in the future.

Finally, banks may decide to spin off certain of their input operations. An example may be the back-office operations or the transactions segment. There is no reason why there should not develop in the future firms which specialize in performing these tasks and sell them to banks. Telecommunications firms, for instance could offer financial transactions services to banks. Data processing firms could offer backroom operations. Telematics again will be the driving force behind such developments.

8. Conclusion

This paper has offered two conjectures regarding the effects of telematics on banking. The first conjecture is that the inter-nationalisation of financial markets would not have been possible without these new technologies and that these technologies will continue to be the driving force in the finance industry. The second conjecture is that in banking the international division of labor will take the form of large multinationally integrated entities competing against each other with more and more specialized products. Due to the multinationalization of banks there will be almost no trade between nations in the field of banking.

These conclusions stand in contrast to the overwhelming part of the literature which usually believes that other factors than telecommunications have been (and are) more important for the internationalisation of financial markets and that banking exhibits a trend towards institutions that offer all financial services under one roof ("Allfinanz").

---

8 Clearly this financial product would be impossible without data-processing and telecommunications. The sheer amount of calculations to be performed during a single business day would amount to an insurmountable obstacle.

References

Arrow, K.J. (1962),"Economic Welfare and the Allocation of Resources for Invention", in *The Rate and Direction of Inventive Activity: Economic and Social Factors*, Universities-NBER Conference Series, Princeton, pp. 609-625.

Barras, R. (1988), "Interactive Innovation in Financial and Business Services: The Vanguard of the Service Revolution" SPRU, University of Sussex, June 1988, mimeographed.

Berlin,M.,Philips,A. (1985), "Technology and Financial Services: Regulatory Problems in a Deregulated Environment", Fishman-Davidson Center for the Study of the Services Sector, The Wharton School, University of Pennsylvania, *Discussion Paper No.8.*

Bryant, R. (1987), *International Financial Intermediation*, Washington D.C., The Brookings Institution.

Coase, R. (1937), "The Nature of the Firm", *Economica (4)*, November 1937, pp. 386-405.

Coase, R. (1960), "The Problem of Social Cost", *Journal of Law and Economics*, 3, October, pp. 1-44.

Dahlmann, C. (1979), "The Problem of Externality", *Journal of Law and Economics*, 22, October, pp. 141-162.

Dennig, U. (1987), *Die Euro-Teilmärkte*, Hamburg, Verlag Weltarchiv.

Dunning, J.(1977): "Trade, Location of Economic Activity and the MNE: A Search for an Eclectic Approach", in B.Ohlin et al. (eds.) , *The International Allocation of Economic Activity*, New York, pp. 395-418.

Glüder, G. (1988), *Die Entstehung multinationaler Banken*, Wiesbaden, Gabler Verlag.

Hymer, S. (1960/1976), *The International Operations of National Firms: A Study of Direct Foreign Investment*, Cambridge, MIT-Press.

Kindleberger, C.P. (1984), "International Banks and International Business in Historical Perspective", in C.P. Kindleberger: *Multinational Excursions*, Cambridge, MIT-Press.

Kogut, B. (1983), "Foreign Direct Investment as a Sequential Process", in, C.Kindleberger/D.Audretsch (eds.), *The Multinational Corporation in the 1980s*, Cambridge/Mass., 1983, 37-55.

Levich, R. (1987), "Financial Innovations in International Financial Markets, National Bureau of Economic Research", *Working Paper* No.2277

OECD (1983), *The Internationalization of Banking - The Policy Issues*, Paris.

Schmalensee, R. (1988), "Industrial Economics: An Overview", *The Economic Journal*, 98, pp. 643-681.

Smith, A. (1776), *An Inquiry Into the Nature and the Causes of the Wealth of Nations*, London, Routledge & Sons, no year

Stigler, G. (1951), "The Division of Labor is Limited by the Extent of the Market", *Journal of Political Economy*. LIX, pp. 185- 193.

Stigler, G. (1968), *The Organization of Industry*, Chicago, Chicago University Press, Chapter 12.

Teece, D. (1985), "Multinational Enterprise, Internal Governance, and Industrial Organization", *American Economic Review*, (75) May 1985, pp. 233-238.

Weizsäcker, C.C. von (1988), "Das Auto, der Kühlschrank, das Telefon", *ORDO, Jahrbuch für die Ordnung von Wirtschaft und Gesellschaft*, Band 39, P.169-171.

Williamson, O. (1985), *The Economic Institutions of Capitalism*, New York, Free Press.

Comment Chapter 7

David Allen

Center for Technology, Policy and Industrial Development, MIT
Cambridge, Massachusetts

First, let's make a moment to appreciate the contribution that Bernhard has offered us.

He takes us back to Adam Smith's basic, compelling argument that "the division of labor is limited by the extent of the market". He prepares us for a modern version of this theory with the corollary insight that extent of the market itself depends on the available transport technology - in our case, technology for *information* transport. Then he brings the theory up to date through Coase, Stigler and Williamson, finally concluding with his own synthesis.

By this, he moves to focus economists' attention on the question of how telematics (in Bernhard's own definition, these are the information and communications technologies) impact industrial organization.

This is done by artfully interweaving into the presentation a case study and empirical data, as well as a review of a companion current debate. The real world tour takes as its subject - and thereby explicates - the farflung, arcane and complex phenomenon, telecommunications networking for international banking.

The product that results is a pair of hypotheses that run counter the fashion in current thinking about international banking. They command our consideration.

And all in twenty pages or so. In my view anyway, a special and elegant treat, indeed.

The Discussion

What is the essential structure of Bernhard's theoretical proposition? Asking his pardon in advance for over-simplification, and at the risk of misinterpreting, I believe the essential structure to be this: Telecommunications reduces transaction costs, and reduced transaction costs encourage the possibility for division of labor, or spinoff. From there of course, he draws conclusions abourt spinoffs/industrial structure in banking.

This abbreviated characterization does injustice to the nuance of Bernhard's story. So how might we take fuller advantage of that nuance? I think the richness of the presentation offers numerous possible avenues to explore. I will suggest one that particularly catches my eye, and appeals to my own predilections. But I caution that thereby I must leave, for another day, the several *other* avenues of interest which Bernhard opens.

From the beginnig of the paper Bernhard leads us to consider division of labor *among* firms, alongside division within the firm. By the time he reaches the theory finale, he is synthesizing three conditions into one, which taken together speak to this question of division of labor among firms. To my eye he is encouraging us to consider the boundary question, in a general sort of way.

To move toward a conception of division of labor *among* firms was a bit of an innovation, as I believe we may instinctively feel as we read it. The point becomes particularly clear when we consider questions of competition, or coordination, in this light.

Boundaries for Competition and Coordination

Again to over-simplify, the usual approach was (and in some quarters still is) to expect competition to occur only among firms, and coordination to occur only within the firm. We all know, however, in reality there is actually plenty of competition within the firm. As Bernhard describes, George Stigler even reminded us that this is the case as he built his 1951 proposition about spinoff and division among firms.[1]

So what of the other counter-possibility, namely coordination occuring ***outside*** the confines of the firm, that is among firms? The inevitable concomitant to the subject of division of labor was, and is, the question of coordinating that newly-divided labor. So to speak of division of labor among ***firms*** invites consideration willy-nilly of coordination among firms.

I suggested that Bernhard encouraged general consideration of boundary questions. What did I mean? If competition and coordination both occur both inside and outside the firm, the firm's boundary is not so much the idiosyncratic, inviolable barrier we have previously been conceiving. Instead, the boundary that matters is the boundary between activities. Yes, of course the firm boundary ***is*** different in some important way from the boundaries which only divide activities. But in terms of outcomes that matter ***economically***, it is the ***activity*** boundary which contains and capsulizes the crucial competition/coordination event. And an activity may encompass units smaller than, or larger than, a firm.

Let's immediately consider the obvious difficulties. US-style competitive marketplace practice *prima facie* strictly forbids coordination among firms. But if we look at other industrialized capitalist economies we find significant variation on this practice. Japan is the case most commonly cited for coordination among firms (and my own limited experience there suggests an intensity of rivalry among Japanese firms which surely guarantees fierce competition domestically in Japan, at the same time). Though I feel personally on less informed ground as far as Europe is concerned, I would ask my European colleagues whether they also see instances of coordination among firms in Europe. Certainly, Americans interested in HDTV think they see some coordination, even coordination which includes government partnership!

Of course I hasten to hold Bernhard blameless for this extension of his exposition; this has been my doing, for better or for worse. However, if these are useful ideas, they strongly suggest new territory for economists to explore, and Bernhard's paper has opened the way. As I can conceive this new territory, one piece to develop is a robust approach to the boundary question. How, for instance, can we reliably identify the unit

---

1 In the section quoted, Stigler mentions for instance internal technological revalry. Can we reasonably imagine competition among units within say a bank, competition, that is, which is similar to the competition we envision among firms competing for the same consumer in a marketplace? Depending on the organizational climate, I expect that that is possible and will be found in at least some instances.

where the competition/coordination event occurs? Is there some phenomenon of an expanding/contracting boundary? In that case sometimes the business unit is the relevant focus, sometimes the whole firm is, other times the full industry.

To raise the boundary question seems also to raise the question whether important cultural differences will be found. To return to the US case, the company boundary in the US seems strong and relatively inviolable. But in Japan - and apparently also Italy as at least one other case - sub-contracting between large firms and smaller, relatively more indigenous firms seems to suggest a different approach to the question of where to establish boundaries, and how to link across those boundaries.

More urgently, the other new piece of territory to explore poses the question: how can we effectively conceptualize competition and coordination occuring together, in the same unit? Is it adequate to suggest a balance of coordination costs against transaction costs, but now both taking place within the same unit? Would some new approach be more helpful? Since the whole paradigm has been predicated on straight competition, does this suggest(surely, all too immodestly) some fundamental rethinking, with a more complicated mechanism put in place to underpin the basis for economic interaction?

What is the role of telematics in this new scenario? In fact, one of Bernhard's main intents, and results, is to focus our attention on this question.

The Impact of Telematics on Industrial Organization

For me at least, one item of anecdotal information from the case study cements Bernhard's contention that telematics was a necessary condition for the multinational expansion of banking. If a failure of the telematics in fact forebodes failure of the banking firm, within no more than a few days, then surely telematics is a basic essential.

To ask the larger question, however - to ask, that is, for the impact of telematics on industrial organization - is to invoke a hoary debate. That debate might be called the question of technological determinism: in its rawest form, this question asks whether technology deterministically constrains, even directs, human outcomes.

Bernhard has served up a convincing argument from an economist's perspective. Lower communications costs (and I might expand this to "increased performance, which is reflected in lower unit communication costs") mean lower transaction costs, and these translate into an industrial structure for banking.

I think, however, most of us would agree that the question of *causation* itself remains an open one, a question that the paper did not attend to address. Nor would many of us be likely to propose such an answer. The situation is surely rather complex, with technical tools and human choice interplaying between each other in subtle, hard-to-describe ways.

What we *can* notice follows on from the questions raised in the prior section. Communications technology surely plays some causal role in both competition and coordination. Adequate comparative information downstream in the value chain is thought, as a standard proposition, to be useful for competion. But on the contrary, whether communication between *competing* units might in some cases also enhance competition has not been investigated so often. Likewise, communication is surely the stuff of coordination. What is less familiar is how communications technologies also may have the capacity to *disrupt* personal exchange. Increased reliance on the telephone, for instance, may have the effect of diminishing the number of face-to-face meeting, with a loss of some understanding between the parties.

More generally, communications technology surely is an important basis for making the linkage between units, once boundaries have been defined. As Bernhard notes in the case study, the bank's telecommunications network was established in conformity with its organizational structure. Perhaps this suggests one useful approach to understanding the role of telematics in structure.

It seems likely that telecommunications networks offer some physical parallel to the human relationships, the organization, which the network supports. For instance, levels of privacy, or publicness, in the network may parallel what is inside and what is outside the various unit boundaries. Such are the beginnings of perhaps one relationship between network and structure.

What are the strong conclusions Bernhard draws about the structure of the international banking industry?

The Impact on Banking Itself

Division of labor will not occur on a locational basis - there will not be national competition on the international stage. But there will be a functional division of labor - large international banks will increasingly specialize their activities. These are the propositions.

I, for one, am satisfied with the first of these arguments. (And I notice that Bernhard here modifies, if I understand correctly, his basic proposition about telematics, transaction costs and the effect on structure. Here he invokes some of the thinking about MNE's and about banks' special use of information.) I find the second argument the most stimulating. Will international banking indeed specialize functionally? On that, much argument turns.

If we follow some of the thinking above, we might add these additional queries:
Banks traffic in money going in one direction, and in information going in the other. Money, on the one hand, is the ultimate commodity. For information, on the other hand, we might take a leaf from Claude Shannon the information theorist, who defined information as that which gives surprise. In this view, information would be the very antithesis of a commodity - for any unit of information to exist, that unit must necessarily be different from any of its brethren.

Banks, then, find themselves with some contradictory basic requirements. Commodities lend themselves to a regime of relatively low transaction costs. But just the reverse is the case for specialized, non-commodity items. Now we apply Bernhard's proposition that lowered transaction costs encourage spinoff. The money side of the bank's activities, where transaction costs may be relatively lower, would tend to disaggregation. The information side would be just the reverse.

This latter is, of course, just the basis Bernhard used to draw his conclusion about multinationalization of international banking (what I have described as the "no locational spinoff" conclusion). Does the former position, about the money side of the banking business, give us the other half? That would be the tendency to disaggregate and spinoff. This, of course, Bernhard has also argued, as spinoffs in a functional direction.

Part IV

Regulatory Models and Policies

8

Changing Regulatory Policies: The Danish Case

Ole Jess Olsen

Institute of Economics
and Planning,
Roskilde University
Roskilde

1. Introduction[1]

This paper deals with the changing regulatory system for telecommunications in Denmark. During the eighties the rules for statutory public monopolies, which were established long time ago, have been the object of extensive modifications in most industrialised countries. The Danish telecommunications industry provides a national example of this process of redefining regulation of public enterprise.

Until 1986 there did not exist a unified system of regulation of the Danish telecommunications industry. The industry was operated as an integrated national system by the Danish PTT *(Post & Telegrafvæsenet)* and three independent regional companies. They belonged to partly separated regulatory bodies among which the division of authority was ambiguous. The approach to regulation was more to persuade

[1] I wish to thank Mogens Kühn Pedersen and Soren Tengvad for helpful comments.

and coordinate than to exercise straightforward authority. The system could be characterised as self-governing with the government acting as an arbitrator.

The fast speed of technological development, not least the integration of voice and non-voice services, made the established organisation and regulatory tradition of telecommunications obsolete. In most countries measures have been taken to change the regulatory system, mainly through liberalisations of the former legal monopolies of the telecommunications administrations. In Western Europe this policy has first of all been advocated by the Commission of the European Communities which considers telecommunications as an important element of the construction of an internal market.

In 1986 a new division of activities within the Danish telecommunications industry was established to handle the new situation. All regulatory powers were vested in the *General Directorate of the PPT* and separated from operations. Shortly afterwards the General Directorate adopted the guidelines stated by the *European Commission.*

These policy lines are presented in the *Green Paper* from June 1987 and specified on later occasions. They include:

- A clear separation of regulatory and operational functions.
- A narrow monopoly confined to the network infrastructure and traditional voice services.
- Unlimited competition in all other services and in terminal equipment.
- Cost-based tariffs which are compatible with the financial viability of the telecommunications administrations.
- Free access for the administrations to the competitive markets on equal terms with other service operators.

It is the object of this paper to discuss our new regulatory policy. First I present the organisational and regulatory tradition of Danish telecommunications. Then I go through the different measures introduced by the General Directorate of the PTT to handle the new situation. How do these measures compare with the proposals for an integrated, competitive market in Europe for telecommunications services? How do they compare with theoretical standards for efficiency regulation? And finally, what are the comparative advantages and disadvantages of Danish telecommunications facing a common market in this industry?

2. Organisation and Regulatory Tradition of the Danish Telecommunications Industry

For a small country with only five million inhabitants, the Danish telecommunications industry has had a convoluted past. Since 1897 telecommunications was a legal monopoly in Denmark with concessionary companies in charge of regional operations and the PTT in charge of interregional and international calls (and some regional operations as well). The network infrastructure was operated jointly by the concessionary companies and the PTT. There was no unified system of control. The concessionary companies were regulated by the Department of Public Works. It had no authority over the PTT which was placed directly under the authority of the Minister of Public Works[2].

After a period this division of labour came into conflict with the potential economies of scale and scope made possible by technological development. The fusion of the existing telecommunications administrations into one national corporation became a logical solution and was in fact suggested by the reform committees which from time to other was set up to solve the problems. But it never became a politically viable solution. Instead a unified national system was reached through persuading the concessionary companies and the PTT to cooperate. An agreement on principles and behavioural rules for this cooperation was signed between the telecommunications administrations and the government in 1950 (called the *Concordate*).

This arrangement, which worked for 35 years, was monitored through coordinating committees established by the telecommunications administrations to handle the many issues of common use of equipment, traffic planning, as well as of prices and cost-sharing schemes. The government participated as an arbitrator. The main deterrent against irresponsible behaviour was a repeated threat of reorganising the sector into one public corporation (Moesgaard Andersen, 1986a).

Specified economic principles such as marginal cost pricing, rate of return regulation, and price caps were never used by the regulatory authorities. This does not mean, however, that it was unregulated. But in accordance with the cooperative framework regulation mainly took place through implementation of agreed policy lines. The concessionary companies were expected to turn up with modest profits and the telecommunications activities of the PTT were expected to contribute to the State

2 The PTT has recently been transferred to the Ministry of Communications and Religious Affairs.

budget. Accounting principles were specified by the Ministry of Public Works. Wage increases should be approved by the Ministry of Finance.

Tariffs were based on a general understanding of the politically acceptable increases combined with a large number of detailed agreements on the sharing of costs and revenues among the different telecommunications administrations. Tariffs were calculated on an average cost-basis and applying a top-down approach. First the increase of revenues necessary for the telecommunications administrations to break even was calculated and accepted by the Ministry. This implied an average mark-up on present tariffs. This mark-up was then distributed among the different services according to not specified considerations of cost increases, political sensitivity, etc. Individual tariffs were not subject to direct approval. And only on very few occasions were the revenue forecast presented by the telecommunications administrations rejected by the Ministry.

The fast speed of technological development, not least the many new services made possible by the integration of voice and non-voice communication, increased as elsewhere the pressure on the established organisation in Denmark. New issues were taken up on an ad hoc basis in the network of joint committees. As these issues multiplied, it became more difficult to reach consensus and update the numerous agreements between the telecommunications administrations on the sharing of equipment, costs, and revenues in due time. Awkward cost-structures created incentives for breaking with cooperation[3]. Customer relations also became more complicated and without any appealing logic (new services like data communications and mobile telephones were operated by the PTT). The old system could no longer be trusted to handle the fast changing conditions of the telecommunications industry.

A government commission was set up with the participation of the telecommunications administrations (White Paper, 1985). Its proposals became implemented during 1985-87. The existing company structure was preserved. A unified system of control was created by vesting all regulatory powers in the General Directorate of the PTT. A special state enterprise has been set up to handle type approval of terminal equipment. In exchange for being exposed to more centralised control, the regional companies got most of the new services, most costumer relations and part of the network infrastructure from the PTT (which has delegated its remaining operational functions to a new state enterprise, see figure 1). In this way the regulatory and operational functions became separated.

[3] It was a violation of the spirit of the "Concordate", when in the early eighties both the PTT and the regional companies started constructing parallel optic fibre transmission lines.

Figure 1: The organisational structure of Danish telecommunications in 1989.

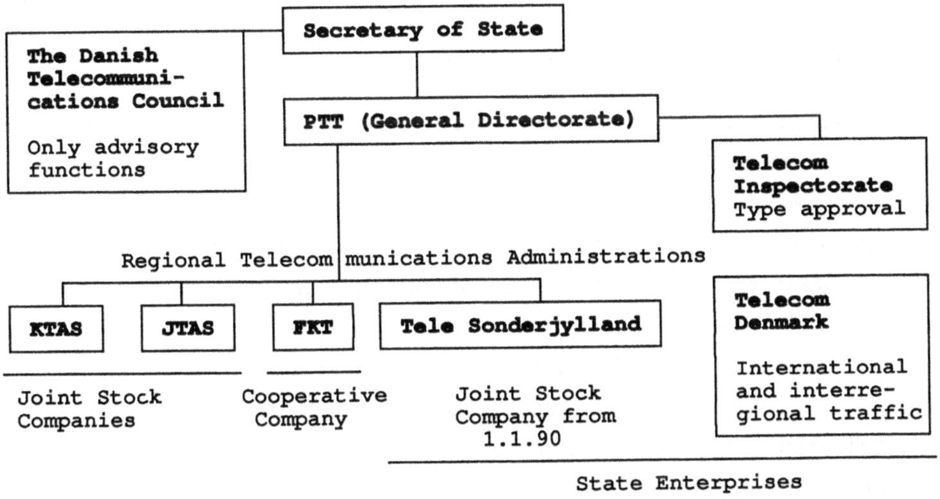

It could be feared that the informal way of handling regulation and the absence of monitoring based on explicitly stated principles would lead to a situation, where the telecommunications administrations got the best from two worlds: A protected monopoly without too much government interference. This was, however, not the case. When comparisons are made with other West European countries, the Danish telecommunications industry evinced a relatively good performance. This characteristic refers primarily to the extension, quality and prices of traditional voice telephony, but can also be extended to include other services as well (Olsen, 1989).

3. Present Development of Regulatory Policies

The General Directorate has extensive regulatory powers over the telecommunications administrations. One of these are under its direct command inside the PTT. And the General Directorate is both the majority shareholder and concessionary authority of the other four enterprises[4]. In principle the PTT is therefore entitled to control whatever it

[4] Usually, it is the concessionary or political authority which is employed in practice. Despite being

wants to control: Prices, financial activities, technical matters, size of staff and wages. Formally speaking, the new organisational structure complies with the separation of regulatory and operational functions wanted by the European Commission. But it is more important to know what this means in practice. How does the General Directorate accomplish its new task as regulatory authority?

A continuation of the former cooperative approach to regulation was a precondition of the political agreement on which the reorganisation was based. So far the regulatory approach chosen by the General Directorate has respected this requirement. It does not intervene unduly into the affairs of the telecommunications administrations and decisions are mostly based on consensus and not on authority (see Würtzen and Abild Andersen, 1987). This choice of continuity was also indicated through the elaboration of new concessions in 1986. Very little was changed in the text from the previous concessions. It articulates traditional regulation of a statutory monopoly, the major object of which is to prevent the abuse of monopoly rights. There is very little in this text anticipating the specific problems of the future regulation in an open European market.

But as already mentioned, it has not been the practice in Denmark to use the extensive regulatory powers specified in the concessions to intervene in the day-to-day business of the telecommunications enterprises. And there are presently no signs of a more activist approach to be chosen by the General Directorate. This characteristic excludes also the elaboration of independent policies like the dirigist French industrial policy and the Anglo-American policy of indiscriminate liberalisations. So far the General Directorate has taken the role as the arbitrator between the European Commission, the Danish political system, the interest organisations (representing private industry, trade unions, and consumers), and the telecommunications administrations. The objects of the new regulatory policy is stated in general and rather vague terms: To continue and develop an integrated national infrastructure, to provide reliable services, on equal terms, with a quality comparable to those of other industrialised countries, and with lower prices (P&T, 1988b). Liberalisations should follow the indications and pace set by the European Commission, no less and no more.

---

a majority shareholder in the two joint-stock companies, JTAS and KTAS, the government only possess a minority of the seats on the board of directors of these two companies. Adding the representatives of the employees and the private shareholders makes a majority. The Government has, however, on several occasions proved willing to overrule this majority. The chairmen of the two companies, who are members of Parliament, serve as liaison officers in such situations.

In the following sections I will discuss the regulatory policies introduced with respect to the different items mentioned in the introduction.

3.1. Liberalisation of Terminal Equipment

The European Commission has since long insisted on a total liberalisation of terminal equipment and finally specified its will in *Commission Directives*, issued in 1988 and 1989 (applying article 90A, which under certain circumstances permits the Commission to issue directives directly. It implies a more speedy implementation than alternative procedures which require the consent of the Council. See the chapter by J. Scherer in this volume). Such directives must be respected by the member states). The Commission has also suggested a common procedure for type approvals and for the elaboration of standards (to be placed with the *European Telecommunications Standards Institute*, ETSI). The idea is that type approvals registered in one country automatically are valid in all member countries. This is to avoid single countries using national provisions as measures to protect national manufacturers and to continue the status of the telecommunications administrations as exclusive dealers.

Denmark has been rather late to start liberalisations of terminal equipment. According to a survey from 1987, we were among the three countries in Western Europe with most items of exclusive provision (see Foreman-Peck and Müller, 1988). The Danish authorities declared the *Commission Directive* of May 1988 to be respected by a timetable for liberalisations put forward by the PTT in February 1988. It was based on a compromise with the telecommunications administations and the business organisations as well. The Danish proposal was, however, not accepted by the Commission. There were two reasons for this refusal. Firstly, liberalisations of the most used terminal equipment, the first telephone set, private exchanges (PBXs), and internal networks, were delayed to the very last moment, January 1992. All other types of terminal equipment have been given free by now, and the Danish authorities have accepted to advance the liberalisation of the remaining items of exclusive provision.

Secondly, the *Danish law* on terminal equipment from 1986 required a procedure of type approval which was not perceived as compatible with the Directive. The law prescribed type approval by the PTT and endowed the Danish authorities with special rights to control dealers of terminal equipment. This was to ensure technical compatibility, but was perceived by the Commission as a regulation of marketing, which is not tolerated. In the new issue of the *Directive on terminal equipment* from July 1989 (KOM(89) 289) it appears, however, as if the Danish position has been accepted.

3.2. Liberalisations of Telecommunications Services

The Commission's concept of the future is one of a narrow monopoly. The telecommunications administrations will continue to be the exclusive providers of the network infrastructure. Among services only voice telephony, which is considered a *basic service*, and telex, which is expected to vanish, will be retained as legal monopolies. All other services will in principle be open for competition (recently also mobile telephones and public paging have been considered for future exclusive provision. These two services are of a special significance, because they represent an alternative to the local cable-network which has hitherto been considered a sustainable monopoly). This includes both the use of the public switched network and of leased lines. And there will be no restrictions on the use of the network, whether being for pure resale of idle capacity or for the supply of value-added services.

The Danish restrictions on the use of the telecommunications network were extensive until very recently (see Commission of the European Communities, 1987: 60). They included prohibition of resale of hired capacity and required special permissions to offer value-added services[5]. Since the beginning of 1988 the speed of liberalisation has, however, been very fast. As the text and data services are concerned, there are no longer any restrictions on the use of the public switched networks. The same conditions apply to the use of leased lines in combination with the public switched text and data networks. The provision of value-added services on leased lines, and in combination with the public switched network, has also been given free. And in addition pure resale of traffic is allowed up to 5 percent of total traffic. In 1989 these provisions have been extended to international traffic on leased lines (see P&T, 1988a).

3.3. Partial Regulation of the Telecommunications Administrations

In the future the European telecommunications administrations (due to the recent privatisations these can be both private and public enterprises) are expected to perform two different roles: As *monopoly providers* of the network infrastructure and of basic services (voice telephony), and as *competing firms* in the fast expanding markets for terminal equipment and new telecommunications services. One of the major regulatory tasks of both the European and the national authorities will concern the proper separation of monopoly and competitive activities. Future regulation is expected to accomplish the following objectives:

---

5 A such permission has since long been issued to private service operators conveying telex from America to other Western European countries. This is an example of abitrage due to the very low Danish tariffs of this service.

- Prevent the telecommunications administrations from using income from the monopoly areas to undercut competitors in the liberalised areas (*cross-subsidisation* and *predatory pricing*) or in other ways to impede efficient competition.

- To protect consumers of monopoly services against misuse of the monopoly.

- To support the financial viability of the telecommunications administrations to make them capable of meeting their special obligations as common carriers and providers of basic services.

- To protect the telecommunications administrations against inefficient competition (so-called cream-skimming[6]).

It seems obvious to me that regulation of telecommunications will be a far more complicated task in the future than it was in the past. Experiences from Japan and the United States, which have already tried partial regulation for some time, confirm this expectation. Which measures have so far been taken by the Danish authorities to cope with these new problems?

The approach to partial regulation chosen by the General Directorate is to assign competitive activities to subsidiaries of the telecommunications administrations or to divisions with separate accounts. Such subsidiaries can be owned exclusively by one of the administrations, jointly by several administrations, or it can take the form of a partnership with private companies. The regional administrations are at present very busy seeking partners for joint ventures. KTAS has established a joint venture with IBM, and the other three regional enterprises have established one with Digital and Siemens to sell data services to business firms.

All capital inputs into such subsidiaries and divisions must be financed by means provided on market conditions. No restrictions on competition must be introduced (for instance obligations to use specific brands of equipment). The General Directorate considers, but has not yet decided, if service obligations should be imposed on the regional administrations in selected fields (like in West Germany, see Federal Minister of Posts and Telecommunications, 1988).

So far the General Directorate has elaborated rules for the following activities: Mobile telephone sets, alarm service, value-added services, telephone directories, and cable-TV.

6 *Cream-skimming* refers to a situation, where a competitor without obligation to serve the whole market undercuts the incumbent service provider in the more lucrative parts of the market.

To illustrate the problems involved in partial regulation, I will discuss three of these cases. Two of them are services (alarm service and cable-TV) whereas the third represents competition in terminal equipment (mobile telephones). Some of the problems indicated are general, others are specific for the Danish context.

Alarm *service* became an issue in the mid-eighties, because the Copenhagen *Telecommunications Administration* (KTAS) started to offer services in competition with the dominating Danish company in this field, the ISS. The alarm service offered by the ISS employed special equipment installed by this company in the public switched network. It therefore complained to the Government that the KTAS could misuse the knowledge acquired as provider of the network infrastructure and thereby expose ISS to unfair competition. The issue was solved by prohibiting the telecommunications administrations to offer certain types of alarm service (surveillance of private firms against burglaries, which is now a regulated business in Denmark). It was also decided to offer access on equal terms to other alarm companies than the ISS and to withdraw the possibility of installing private equipment in the public switched network. Other alarm services such as surveillance of public institutions and technical installations are, however, open for the telecommunications administrations.

The underlying efficiency questions were never addressed: If alarm service is best provided in a competitive setting or as a regulated monopoly, and if it is more efficient to integrate the services in question into a telecommunications firm or into an alarm service firm (where are the best economies of scope, see Neumann, 1987). It is not unlikely that it will be more economical in the future to install the equipment for some value-added services directly in the public switched network rather than to attach it as terminal equipment. But how are the monopoly activities then separated from the competitive activities in a meaningful way and how is the most efficient solution selected?

Mobile telephone sets can be used to illustrate problems common to a competitive market for terminal equipment. The Nordic countries introduced some years ago a very successful mobile telephone system which was the first in the world to reach a nationwide coverage[7]. Before 1986 the system was administered by the PTT as a national service and the telephone sets were offered by different manufacturers on a competitive basis. The regional administrations offered one among several brands which soon gained a significant share of the market. The decision to transfer all costumer related activities to the regional administrations included also the mobile telephone

[7] Denmark reached 100.000 subscribers in December 1988.

service. Then the manufacturers of the competing brands started to complain. They argued that under the new conditions the telecommunications administrations would sell both the service and the telephone set, thus favouring the products they offered. To solve the issue a special procedure was introduced which violates the principle of locating all costumer relations in one place. The initial registration will now take place with the Telecom Inspectorate which is not selling telephone sets. After the registration the regional administration will take over.

This is of course a general problem. At present the regional administrations are establishing subsidiaries for terminal equipment. They are free to sell the products they prefer and to marketing it on commercial terms. Under such circumstances it seems impossible or very artificial to avoid the administrations using their position as providers of basic services to promote the sale of their own products. Most likely there will also exist economies of scope in using the same service organisation for the local installations belonging to the network infrastructure and the terminals attached to it.

Cable-TV is the third area to be discussed. Denmark has a long tradition of local cable-TV. About fifty percent of the households belongs to a cable-TV association. In 1985 it was decided to construct a hybrid network (using both analog and digital technology) for the transmission of high quality signals. In this way the regional administrations planned to become the proprietors of a high quality local network which is considered a strategic asset with regard to future competition. Its first mission is to provide TV-programs to the local associations, who pay for the construction of the system. Later on it is expected to serve as the local network in a nationwide *broadband* for integrated two-ways communications. There have been many uncertainties about the service concept to be used. Some political parties suggested the traditional common carrier concept with service obligation and a common price, others wanted more cost-based prices.

The original solution was a hybrid like the tecnical infrastructure. The monopoly of the reception of satellite signals was uphold. The telecommunications administrations were obliged to extent the hybrid network to the entire country. But tariffs should bear some relation to the costs of local connections. The solution was, however, politically unstable and became fiercely criticised in the mass media. As a result, cable-TV became liberalised in 1987 according to the following priciples: Private persons (cable-TV associations) and firms are now allowed to receive satellite programs directly, and therefore have the choice between this option and the service provided by the cable-TV divisions of the telecommunications administrations. Transmission of TV and radio signals across the boundaries between different districts is still a statutory monopoly (Andersen, 1987).

The regional administrations have been relieved of their former obligations to provide services outside the profitable areas, mainly the larger cities. Each of them has created a cable-TV division, who buys the signals at a local exchange. After a slow start the build-up of the market has been quite rapid. About 350.000 households (about 1 million households are members of local cable-TV associations) will be connected to the hybrid network during 1989.

The redefinition of the hybrid network in 1987 also included a specification of targets for future broadband services for business firms, thus emphasising the original intentions of *integrating broadcasting and telecommunications.* From the beginning of 1992 the telecommunications administrations are obliged to satisfy every demand for broadband services within three month. Tariffs will be equal in all parts of the country, except for the local costs of connection.

Cable-TV is the first important example of separate accounting of monopoly and competitive activities. But a large part of the network used to transmit the TV-signals is common with other services. And it is in no way guaranteed by the accounting procedure applied that the allocation of costs between the regional administrations and their cable-TV subsidiaries is efficient, nor that it is fair.

3.4 Tariffs

The general principles suggested by the European Commission for the future tariffs say that they should be cost-oriented, be properly published and apply to all users on a non-discriminatory basis. It is common to claim that present tariffs do not reflect the underlying cost structure. Consumers are charged the same price for being connected to the public network despite of large variances in the real costs of installation. Subscription charges often discriminate between business and residential consumers. Usage sensitive tariffs vary with distance in a way which has no relation to actual costs. The two other principles quoted above are part of the traditional public service obligation imposed on the telecommunications administrations. Their meaning will, however, be different in a future setting with competition and partial regulation.

As mentioned above, Danish tariffs did not differ significantly from this general picture. The division of the telephone bill between installation, subscription and usage charges was arbitrary. The charge for trunk calls increased proportionally with distance.

Discrimination between business and residential consumers has not been practiced. Costs were distributed on the different items of the telephone bill according to average costs and to arbitrarily chosen principles of allocation. This does not mean, however, that the Danish telecommunications administrations were not concerned with the allocation of costs and its relation to tariffs. The organisational structure with several administrations running an integrated national network made cost-sharing a constant issue of conflict[8].

Since the reorganisation the General Directorate of the PTT has attempted to introduce more transparent principles for the calculation of telecommunications tariffs. So far this has resulted in the announcement of some general principles concerning the overall level of tariffs (P&T, 1988b):

- Increases for a longer period must be 2-3 percent below the increases in consumer prices.
- Tariffs must stay below the level of tariffs in other West European countries.
- Availability and quality of services must keep pace with the demand of the consumers and with the opportunities opened elsewhere.
- This must be compatible with the financial viability of the telecommunications administrations which are obliged to pay most of their investments by the proceeds from their sales.

Productivity growth is now quantified by the PTT. Quality targets are elaborated for access to the public switched network. But it is not very clear how the principles are translated into the annual approval of tariff increases (strictly speaking it is not increases of tariffs but increases of revenues which are approved). Neither is it transparent after which criteria individual tariffs are decided. Due to the introduction of partial competition this is a very delicate problem. The traditional tariff structure in Denmark includes, relatively to other West European countries, high access charges, average subscription fees, and relatively low usage charges. This observation is both relevant for voice service and for non-voice services (see BEUC, 1988; and Eurodata, 1988).

---

8 That the structure of tariffs were out of order with the structure of costs became clear in 1986, when the terms of transfer of services from the PTT to the regional administrations were decided. The latter were compelled to pay 4 percent annually of their future revenues from the monopoly services as a compensation to the PTT for lost income (White Paper, 1988: chapt. 13). This is an estimate of the extent to which trunk calls and international calls previously have been overpriced.

An important question concerning the emerging liberalised system is the relative pricing of different parts of the network which can be used as alternative channels of communication. Examples are alternative local networks for telephones and cable-TV, and the choice between leased lines and the public switched network. Some tariff structures will create incentives for bypass which can be harmful both to efficiency and to the financial viability of the telecommunications administrations. The use of leased lines under competitive conditions is the first concrete area which has been taken up by the European Commission to break the way for the Open Network Provision (The Commission's concept of the future internal market for telecommunications services, GAP, 1989).

Up to now charges for leased lines have been divided into an installation charge and a subscription fee. Usage sensitive charges have not been used. Compared to the public switched network the level of tariffs have been relatively low. This means that there exist strong incentives to move traffic from the public switched network to leased lines. To avoid such consequences under the present more liberal conditions, the Danish telecommunications administrations have started to implement a new structure of tariffs. Its most important feature is a marked decrease in the dependance on distance. The accompanying changes of tariffs (short lines will be more expensive, long lines cheaper) are expected to rise proceeds from leased lines by about 60 percent (under the assumption of a constant pattern of demand), thus making them a less attractive alternative in proportion to the public switched network. The level chosen is estimated to equal the price of three hours of use of the public switched network. But this is of course not always enough to take away incentives to exploit leased lines for pure resale (therefore the ceiling of five percent for the resale of capacity).

The tariffs of the public switched network are also being changed to make them more fit for future competition. The dependance on distance has already been reduced significantly and next year there will only be three tariff zones[9]. These changes have either been decided independently by the telecommunications administrations or through common negotiations with the General Directorate of the PTT.

[9] Between 1983 and 1988 the KTAS increased its prices for local calls by 30-44 percent, whereas trunk calls were only increased by 7 percent. Installation charges and subscription rentals were increased by 17 percent. The changes introduced by the other regional administrations were of a similar magnitude.

4. Some Theoretical Reflections

The basic idea behind the new regulatory system proposed by the European Commission is that a free and large market is the best devise to secure an efficient supply of telecommunications services. The position taken by the European Commission is, however, not unconditionally in favour of free competition. In the papers published by the Commission it is possible to find references to a number of different policy objectives: Industrial policy, fairness, economic efficiency, financial viability of the telecommunications administrations, and social equality. Unfortunately, these policy goals are not clearly stated and separated, wherefore it sometimes becomes difficult to understand how they apply to the concrete policies recommended by the European Commission (Commission of the European Communities, 1987). In this section I will only discuss issues related to efficiency.

Theoretically it is straightforward to develop arguments demonstrating that it is possible to reach an improvement of the present situation through the establishment of a larger market with more competition. It is also possible to create scenarios with estimates of the welfare gains which are likely to arise from a common and competitive market for telecommunications.

Such argumentation assumes that market imperfections are not too severe. But the telecommunications industry is also in the future expected to exhibit significant market imperfections due to economies of scale and scope (common production of several services), network externalities, etc. According to some scenarios of the expected technological development such imperfections can very well increase and not decrease (for instance due to the substitution of terminal equipment by more intelligent transmission and switching installations). In addition, the argumentation is static. It only demonstrates that the general welfare can be improved by moving from one to another situation. It doesn't solve the dynamic problem of getting from here to there. And this process is indeed very important. It includes the establishment of a new regulatory regime and a number of distributional issues as well. That only winners will be created is highly unlikely. In this paper I will confine the discussion to the regulatory issue.

Presently the European Commission is building a new telecommunications bureaucracy in Brussels with its own rules. But the national regulations of the member states are not going to disappear. We are therefore moving towards a twofold system like the American one. The emerging distribution of roles also carries some similarities to the distribution existing in the United States with the Federal Commission (the European

Commission) pressing for liberalisations and harmonisations, and the state commissions defending their regional service providers (the Baby Bells). It is far from clear to me that a such system unconditionally will lead to less and better regulation. With better regulation I mean that it will satisfy the demand for future telecommunications services in a more efficient way (concerning both technical and allocative efficiency) than alternative systems.

So far the main achievements of the European Commission have been to speed up the process af liberalisation in the member countries and to create a framework for common standards. Possibly also research and development has been stimulated. But when it comes to the elaboration and enforcement of new regulatory principles, the results are not impressive. This concerns in particular the elaboration of tariffs. The sections on tariffs in the Green Paper and in the succeeding reports are stated in general terms and avoid most of the difficult issues.

The first proposal from the Analysis and Forecasting Group (GAP) on the implementation of the Open Network Provision for leased lines validates this impression (1989: Chapt. 7). Repeatedly, it is claimed that tariffs shall be cost-based, that "cross-subsidisation" as well as "cream-skimming" shall be avoided, that fair competition shall be ensured, and that the financial viability of the telecommunications administrations shall be secured. But turning to the concrete proposals of the report very little is said concerning the precise operationalisations of the announced principles.

It is well known that the elaboration of cost-based prices is a difficult task in a setting with large common costs. These problems of allocating costs between different services become even greater in a system like telecommunications due to the existence of network externalities (see Sharkey, 1982: Chapt. 9). If a well-founded methodological basis for cost-allocation is not established by the regulatory authorities, it is utterly arbitrary what is called "cost-based", "cross-subsidisation", or "cream-skimming". This means that the floor is open for all kinds of political pressures. None of the principles discussed in the recent theoretical literature and used in the regulatory practice in some countries are mentioned in the reports from the European Commission (about incentive regulation, see Baron (1989); Braeutigam (1989); and Vickers and Yarrow (1988)).

A related and important issue concerns partial regulation. The telecommunications administrations are both serving markets where they have a monopoly and markets with substantial competition. According to which criteria should costs be allocated between the two types of services to avoid inefficiencies? The problem is particularly severe when significant interdependencies exist in demand or in production. Such interdependances can occur in all parts of the system: Between the network infrastructure and different types of services as well as between the infrastructure and terminal equipment (see above and Neumann, 1987). In cases where the inter-dependencies cause significant economies of scope, imposed criteria of fairness and equal terms can very well result in inefficient competition, because the telecommunications administrations are not allowed to price competitive services according to their social costs. Different regulatory models have been suggested to solve this problem. Price-cap regulation is a prominent example, which has either been introduced or is being considered in a number of countries (among these are the United Kingdom, the United States, Japan, France, and West Germany; see Braeutigam and Panzar, 1989; and Heuermann and Neu, 1988).

Theoretical criteria of economic efficiency (like marginal cost and Ramsey pricing) did never play a role in Danish regulatory practice. This observation also applies to modern incentive regulation. The regulatory criteria, which have been used, are the outcome of a bargaining process, in which both telecommunications administrations, government, political parties, and interest organisations take place. As a result of the reorganisation in 1986 and the pressure from the European development, measures have been taken to introduce more explicit and transparent principles. But the theoretical discussion of cost efficiency and the best incentives for public and regulated enterprises still has no place in Danish regulatory practice. And it is possible to find cases, where Danish practice seems to be contrary to salient principles of incentive regulation.

The coexistence of several telecommunications administrations presents an excellent case for the application of yardstick-competition (see Sappington and Stiglitz, 1987; and Vickers and Yarrow, 1988). The idea is to use the comparative cost information as a proxy for competition to stimulate cost efficiency.

There is no such mechanism working in the Danish regulatory system, at least not explicitly. The annual mark-up of revenues for each of the four regional administrations does not seem to bear any direct relationship to their cost performance. Sometimes administrations exhibiting a bad performance will be rewarded by high increases and sometimes they will not. Despite of this, representatives of the telecommunications administrations claim that yardstick-competition is working in practice. Those who turn up with a bad record are put under pressure from the regulatory authorities, politicians, interest organisations, and the mass media[10]. It is not stipulated explicitly into the form

[10] The regulatory procedure for tariffs starts up with suggestions coming from the

of regulatory constraints, but nevertheless functions as a deterrent. Next year the General Directorate will start publishing comparative productivity measures to make this process a bit more open.

5. Comparative Advantages of Denmark Facing an Internal Market for the Telecommunications Industry

Under the expected future conditions the Danish telecommunications administrations will face competition from private service operators and from the other West European telecommunications administrations as well. How fit are the Danish administrations for coping with this challenge? This is a question both of the relative efficiency of the five telecommunications administrations as well as of the Danish regulatory system.

5.1. Comparative Performance

In a comparative perspective, and despite its apparently clumsy organisation, the Danish telecommunications industry has demonstrated the ability to develop efficient and modern services. Concerning output and quality of traditional voice services, Denmark has the third largest extension of subscribers and the fifth largest number of telephones pr. capital among the industrialised countries. Comparative studies of prices and costs usually place the Danish administrations among the best performing telecommunications administrations. In a survey of different comparative studies the Danish industry was consistently placed among the low-price countries in Western Europe (Moesgaard Andersen, 1986b and 1987). Studies of cost-efficiency demonstrate similar good results for the Danish administrations. This is the case both using simple indicators of labour productivity and more sophisticated measures of total factor productivity for the comparison (Foreman-Peck and Müller, 1988).

Price-cap regulation of tariffs is sometimes related to a general price-index (i.e. the British RPI-X where the retail price-index is applied). The price-cap is then supposed to express the potential of relative cost reductions in the telecommunications industry due to technological development as well as to increases in demand. During the seventies the relative price increases of the Danish telecommunications administrations were very low

telecommunications administrations about their need of revenues for the next year. Before submitting the proposals to the General Directorate, they are discussed by the councils of consumer representatives attached to each administration. After negotiations have taken place, the Directorate will announce an average percentage increase of revenues for each administration. There exists no common procedure for the approval of individual tariffs. When problems arise or new services are introduced, the issues involved will be taken up on an ad hoc basis, either by the General Directorate or by the regional administrations.

(6-7 percent below the retail price-index), whereas the gap has become much smaller in the eighties (P&T, 1988b). There are several explanations of this trend. One is that the productivity gains made possible by the introduction of automatic switching have been exhausted. Another explanation emphasises the large investments in the installation of new equipment and the development of new services which all have to be financed from the annual proceeds of the administrations.

The relatively good performance up to now can of course be considered as a competitive advantage. This is in particular so with respect to the other West European administrations. It is, however, not necessarily the case with respect to the private service operators which are often well-established computer firms with a much better competitive experience than the telecommunications administrations.

It is not so obvious that the above mentioned characteristic of the Danish telecommunications industry can be extended to the new services. Even though the Danish industry is more efficient than the British and probably also the American industry when costs are compared, this is no longer so when it comes to the development of new services which constitute the main growth potential of the telecommunications industry. The Danish administrations have not been as fast as the administrations in these countries to introduce new value-added services. However, the introduction of new services is not unconditionally a measure of dynamic efficiency. In the United States a large part of the new value-added services are used for verification of payments between financial institutions as well as between these and their costumers. Such services are not needed in Denmark because an integrated national system of electronic payments has been established through cooperation between the associations of the Danish banks and saving banks and the telecommunications administrations.

5.2 Regulatory Tradition

Similar observations can be made about the Danish regulatory tradition. It certainly includes qualities which can be transferred to the more competitive environment of the future. Among these I will list pragmatic cooperation among several service operators, and competitive practices applied in procurement. But our tradition also exhibits some problems. These can occur due to the small size of the Danish telecommunications firms, and because the cooperative tradition is too slow and will incur too high transaction costs under more competitive conditions.

Taking the large West European countries as examples, the Danish regulatory tradition is far from both of the two extremes: The French dirigist system with its great ambitions of industrial policy, and the British liberal approach with its encouragement of comprehensive competition. Despite being extremes, the telecommunications industry of both countries have a good record from the development of new services during the eighties.

Dirigist measures have never been the tradition in Denmark due to the opposition of strong interest organisations. The business associations participate actively in the elaboration and administration of industrial policy. This observation also applies to the telecommunications industry, for which reason government does not have much space for independent policies. The business associations are not advocates of far-reaching liberalisation either. They believe that an integrated national system run by the telecommunications administrations is a better assurance of the development of reliable and economic services than outright liberalisation. Representing small firms in a small country they can hope for more influence on the development of telecommunications services from this type of organisation than from alternative organisations which are likely to be dominated by large foreign companies (Industrirådet, 1986).

The procurement policy of the Danish telecommunications administrations represents a competitive advantage facing the internal market. Denmark is without large national manufacturers of telecommunications equipment and vertical integration of manufacturing and services has only occured to a limited extent. Competitive tendering has since long been the normal procurement procedure for the construction of the network infrastructure. And it is the deliberate policy of the telecommunications administrations always to use alternative designs. In this way they both avoid strategic dependance on a single supplier and are able to test different technical solutions. The only restriction is that the equipment must be compatible with the national network. The reputation as both competent and independent purchasers has made it particularly attractive for manufacturers to obtain contracts with the Danish administrations. It means that the manufacturers sometimes will offer lower prices to achieve this purpose.

Before there existed some kind of vertical integration between the administrations and the Danish manufacturers of telephone sets. But anticipating future competition in this field, the telecommunications administrations have loosened their dependance on single manufacturers. This observation also applies to other types of terminal equipment.

5.3 Size and Market Prospects

Telecommunications are usually regarded as a natural monopoly because of the existence of large economies of scale which is an important reason for the existence of statutory monopolies in this industry. Knowledge of the precise economies of scale is, however, very limited (Sharkey, 1982: Chapt. 9), and technological developments are probably changing the underlying cost-structure very rapidly. The proliferation of new services and the introduction of integrated modes of transmission and switching (*broadbands and integrated service digital networks,* ISDN) are also likely to cause significant changes in the relative advantages and disadvantages of common production (economies of scope). To this should be added the many delicate problems of organisation, the cost function of which is not likely to exhibit the same features as those existing for technical operations. And when forces are moving in different directions, it is of course very hard to predict the consequences for the best future organisation of the telecommunications industry with regard to scale, and to joined or disjoined production.

The Danish telecommunications administrations are very small compared to most other administrations in Western Europe, not to speak of the United States. The two largest regional administrations (the JTAS and the KTAS) have slightly more than one million main lines, whereas the smallest one (Tele S nderjylland) has 115.000 main lines. Only the telecommunications administrations of Greece, Ireland, and Luxembourg are of a similar size. The administrations in the large West European countries all have more than 20 million main lines. But the record so far has not been to the disadvantage of the small Danish administrations indicating that some economies of scale can been exploited in relatively small organisations.

A kind interpretation of the Danish development would emphasise the combination of cooperative exploitation of production economies of scale (network infrastructure) with yardstick-competition (performance) among several small telecommunications administrations. It could be argued that this type of organisation will continue to possess qualities due to its flexibility. Activities with significant economies of scale will be exploited jointly, either as a part of the new narrow monopoly, advocated by the European Commission, or through joint ventures in the competitive services. In other activities the small telecommunications administrations can choose to work independently and join with relevant private partners. As mentioned above there are signs indicating that a such pattern of organisation is emerging in marketing of terminal equipment and value-added services.

But there can also be high transaction costs associated with this type of organisation. In an environment where the telecommunications administrations will meet each other, not only through yardstick-competition, but also as direct competitors, the necessary efforts to persuade them to cooperate will probably increase. How will the regional administrations react to other administrations initiating direct competition in their territories? Stronger incentives for non-cooperative behaviour are likely to raise transaction costs.

Recently there have been examples of successes as well as of failures with respect to cooperative creation of new services. The introduction of ***mobile telephones*** is a case where cooperation has succeeded. ***Videotex*** is an example of a failure where the telecommunications administrations have experienced difficulties in cooperating. Up to now they have not been able to agree on a system appealing to a large public despite the fact that Denmark should offer rather good market opportunities for videotex. The latter consideration concerns the many large organisations of distribution existing in both the private sector (retail and wholesale trade, booking) and the public sector (libraries, law registers, land registers). A pilot project was initiated by the telecommunications administrations in 1982. But they could not agree on a simple and cheap solution like the *French Teletel* with appeal to a larger public. Neither could they agree on a strategy for attracting potential suppliers of data files and for combining them with their potential consumers. Today there are only a few thousand subscribers to the system which places us among the West European countries with the lowest extension of this service (Andreasen et al., 1989).

With respect to the future market possibilities of the Danish telecommunications industry, the hardware market (capital goods and terminal equipment) is likely to continue under the dominance of a few international firms. The manufacturing part of the Danish industry is predominantly composed of subsidiaries of these firms. It is very unlikely that the Danish firms will participate in the manufacturing of hardware except for some limited product lines, characterised by small scale operations.

With respect to services, prospects are much better. Markets are more open, large scale is not always important, and economies of scope in joint production with the network operators are more often occuring. The small size of the Danish market can in some respects be considered an advantage. Communications among large companies in large cities is the most lucrative market for both network services and value-added services and thus the most attractive for potential competitors. Denmark has very little of both. The Danish market for telecommunications services is composed of predominantly small, but organised units. Examples are the cooperative housing associations (cable-TV), chains of retail shops, business associations, local public institutions (schools, libraries). Such an environment makes bypass difficult. It is important to be an insider. In a Danish context the telecommunications administrations are large organisations with well-founded relations to different types of consumers and their associations. Therefore, a private service operator will most likely find it a more economic choice establishing a joint venture with these administrations rather than establishing its own subsidiary.

6. Concluding Remarks

The Danish telecommunications industry has a peculiar organisational and regulatory tradition. The reactions to the recent fast changing conditions of the industry were first cautious. A minor adaption of the organisation was introduced to separate operational and regulatory functions. All regulatory powers are now vested in the General Directorate of the PTT. After a slow start liberalisations have been speeded up to satisfy the requirements of the European Commission and to adapt the industry to more competitive conditions. This also includes the introduction of tariffs which better than hitherto reflect the underlying structure of costs.

The Danish context exhibits some of the features proposed by modern incentive regulation. One is the possibility of comparing the performance of different service operators. But the Danish tradition of establishing workable compromises between the relevant actors is far from the clear-cut distinction between a principal (regulatory agency) and an agent (telecommunications administration) which is central to incentive regulation. Theoretically based considerations of economic efficiency has never played any part in the Danish regulatory approach.

The final section of this paper was devoted to the advantages and disadvantages of the Danish telecommunications industry in a liberalised market. Will it be possible to continue with several telecommunications administrations running the network infrastructure on a cooperative basis and at the same time being competitors in value-added services and terminal equipment? Will there be any room left for the small Danish administrations on a market, where the competitors are the large data communications firms and the large administrations from the other West European countries? I didn't end up with simple and conclusive answers to these questions and I don't think they exist. The volatile development of the industry and the many unresolved problems concerning the creation of an European regulatory system renders precise forecasts a futile business.

References

Andersen, J.S. (1987), "Hybridnettet", *Nordisk Administrativt Tidsskrift* (68), 369-82.

Andreasen, J. et al. (1989), "Videotex i Europa", *TDS-rapport* no. 2, Institut for Samfundsfag, Danmarks tekniske Hojskole, Lyngby.

Baron, D.P. (1989), "Design of regulatory mechanisms and institutions", in R. Schmalensee and R. Willig (eds.), *Handbook of industrial economics*, Vol. II, Amsterdam: North-Holland.

BEUC (1988), *Consumers and the telephone services in the EEC*, Dok. 238/87, Brussels.

Braeutigam, R.R. (1989), "Optimal policies for natural monopolies", in R. Schmalensee and R. Willig (eds.), *Handbook of industrial economics*, Vol. II, Amsterdam: North-Holland.

Braeutigam, R.R. and J.C. Panzar (1989), "Diversification incentives under 'price-based' and 'cost-based' regulation", *Paper presented to the E.A.R.I.E. Sixteenth Annual Conference*, Budapest.

Commission of the European Communities (1987), *Green Paper on the development of the common market for telecommunications services*, (COM(87) 290), Brussels.

Eurodata (1988), *Eurodata Foundation Yearbook, Revision of tariffs 1988-89*, London, October 1988.

Federal Minister of Post and Telecommunications (1988), *Reform of the postal and telecommunications system*, Heidelberg: R. v. Decker's Verlag.

Foremann-Peck, J. and J. Müller (eds.) (1988), *European telecommunication organisations*, Baden-Baden: Nomos Verlagsgesellschaft.

GAP (1989), *Proposal by the Analysis and Forecasting Group (GAP) on Open Network Provision (ONP) for leased lines in the Community*, Brussels.

Heuermann, A. and W. Neu, "Ein Regulierungsrahmen für das Unternehmen Deutsche Bundespost Telekom", *Diskussionsbeiträge zur Telekommunikationsforschung* no. 43, WIK, Bad Honnef.

Industrirådet (1986), *Industrien og telesektoren*, Copenhagen.

Moesgaard Andersen, M. (1986a), *Omstillingen er klar*, Copenhagen: Poul Selvig.

Moesgaard Andersen, M. (1986b), "Internationale komparative unders gelser på telefonprisområdet", *Teleteknik* (37), 14-22.

Moesgaard Andersen, M. (1987), "Telefonprissammenligninger - en gordisk knude", *Teleteknik* (38), 90-103.

Neumann, K.-H. (1987), "Models of service competition in telecommunications", *Diskussionsbeiträge zur Telekommunikationsforschung* no. 28, WIK, Bad Honnef.

Olsen, O.J. (1989), "Deregulation and reorganisation: The case of the Danish telecommunications industry", *Annals of Public and Cooperative Economy* (60), no. 2.

P&T (1988a), *Annual Report*, General Directorate, Copenhagen.

P&T (1988b), *Notat om generelle takstændringer for nationale tjenester*, General Directorate, Dok. nr. 79/88, Copenhagen.

Sappington, D.E.M. and J.E. Stiglitz (1987), "Information and regulation", in E.E. Bailey (ed.), *Public regulation*, Cambridge Mass.: The MIT Press.

Sharkey, W.W. (1982), *The theory of natural monopoly*, Cambridge Mass.: Cambridge University Press.

Vickers, J. and G. Yarrow (1988), *Privatization: An economic analysis*, Cambridge Mass.: The MIT Press.

White Paper (1985), "Arbejdsdelingen på telekommunikationsområdet", *Betænkning 1030*, Copenhagen.

Würtzen, H. and J. Abild Andersen (1987), "The new division of activities within the Danish telecommunications sector", *Teleteknik* (38), 11-18.

9

Chances and Problems of Liberalization of Telecommunications in Hungary

Emilia Nyevrikel

Research Institute of the
Hungarian PTT (PKI),
Budapest

Political life and the whole economy are undergoing fundamental changes in Hungary today and this fact influences the future of telecommunications as well. Therefore, in the first part of my presentation I give a short overview on the Hungarian economic reform and in the second part I deal with the situation and future of telecommunications in Hungary.

1. Economic Problems - Legislative Changes in Hungary

The centrally planned economy has been existing since 1949 in Hungary. The first version of the very strictly controlled economy ("command economy") was replaced by a softer one in 1968 which gave more autonomy for the enterprises (the present efforts in Czechoslovakia, Bulgaria and to some extent in Soviet Union can be compared to the concept of the Hungarian reform of 1968). For today, however, it turned out that this system doesn't work and Hungary would run into a crisis unless it implements fundamental changes in political life and economy.

The following brief informations illustrate the economic problems of Hungary today:

- There is an 18 billion US$ national debt (1800 $/capita) in 1988,
- There is a 40 billion Hungarian Forint (HUF) state budget deficit (the total budget amounts to 600 billion HUF in 1988),
- Hungary has a poor international competitiveness,
- An unsound economic structure, i.e. there is a high proportion of not profitable companies, and
- High inflation (15-17%).

The goal of the present economic reform is to shift Hungary from a centrally planned economy into a market economy. This reform has many-sided and substantial effects on political life as well as on society : political parties must withdraw from the economic decisions, the role of the state must be re-evaluated, etc.

Several measures have been taken in the direction of a market economy in the last two years:

1987

The one tiered *bank system* was replaced by a two tiered one. In this system the central bank is responsible for formulating and executing the monetary policy in Hungary, and commercial banks and financial institutions collect deposits and supply the companies and inhabitants with loans. Before 1987 there were only 7 banks in Hungary (with shared markets) and today about 30, including banks with 100% foreign stake.

1988

A new *tax system* was introduced in two areas:

- The jungle of taxes of companies was replaced by two main taxes: the value added tax in which there are 3 tax rates (0%, 15%, 25%; telecom: 0%) and the profit tax (50%)
- A western type personal income tax system was introduced.

Bankruptcy Act. This act determines how to restructure or sell off a bankrupt company. In spite of the existence of this law the elimination of the bankrupt companies is getting on slowly (for fear of unemployment and due to strong industrial lobbies) and 40% of the state expenditures is spent on subsidies of not profitable companies.

A Stock exchange was established. Because of the small market of securities it has meetings once a week today and the market is characterised by domination of bonds.

Foreign Investment Act, which declares:

- foreign investments enjoy full safety,
- in case of nationalisation investments must be compensated immediately in the currency the investment was made,
- profit can be repatriated in the currency the investment was made.[1]

1989

Company Act. Under this act,

- residents can have shares,
- individuals have right to set up private entities,
- under specific circumstances joint stock companies with foreign participation enjoy 60-100% tax discount.

Since the existence of the Company Act 1700 new companies have been set up (total number of companies and cooperatives is 15.000).

Act of transformation of state enterprises. It simplifies the transformation of state enterprises into public limited companies.

Import liberalization. Companies needed an import permission from the Ministry of Trade so far. Import will gradually be liberalised by 1992.

Establishment of a "*Fund of state assets*". This organization would incorporate the owner of state enterprises (collects and reinvests dividend of state shares, etc).

---

1 For further information see the appendix which compares the conditions of foreign investments in East Europe.

Despite the above-mentioned liberalization measures the positive effects of competition are slow to show which is due to the ownership system of the country. We must accept the fact that market economy can not be implemented with the public ownership dominating and the "ownership reform" seems to be the crucial point of the economic reform in Hungary.

In this respect two questions arise:

a) What is the *sound proportion of public/private ownership* in Hungary? There is no consensus on this question. The leftist parties prefer less privatisation, the rightist organisations, however, want to carry out as much privatisation as would be desirable to make private ownership dominant so that private actors determine market behavior.

b) *How to implement* reduction of private ownership. We have to be aware of the fact that the problem is unprecedented: *big volume* of state property must be reduced *alongside a small proportion of private sector* and during a *short period of time.* This problem is of great importance and it will arise in all East-European countries which will join the present reform-course.

Theoretically the following ideas came up in Hungary:

1. Giving the nationalised assets to the *original owners.* This solution arose just theoretically because it is very contradictional: how to state the original value of the properties; owners (or inheritors) have to be found, etc.

2. *Traditional privatisation,* i.e. selling off the public enterprises to the private sector, individual entrepreneurs or foreign investors.

In market economies selling off the nationalised assets is adaptable because there is a substantial private sector which is able to buy the offered assets. In Hungary, however, the division of the economic branches by ownership is the following (Dec. 31., 1987):

	%
State enterprises	77.5
co-operatives	21.0
private sector	1.5
	100.0

The value of the assets of state enterprises and cooperatives is estimated at 2000 billion HUF, while the deposits of individuals amount to 300 billion and a substantial part of these deposits were for the purpose of postponed consumption. The money the residents would spend on investments is estimated at 15-30 billion HUF, which means privatisation through the purchases by residents would also take a very long time.[2]

Despite the fairly liberal investment regulation foreign investors are still distrustful and consider Hungarian investments risky. So far 300 million US$ were invested by foreign entrepreneurs in Hungary.

The purpose of the "Act of transformation of state enterprises" was to support this privatisation, because it's only the limited liability company that can be sold off on the capital market. By this Act 20% stake of the rest can be sold off. 80% of the income from the sold shares has to be transferred to the "Fund of state assets" and 20% can be kept by the company (to stimulate enterprise transformations).

State enterprises aren't forced to transfer into ltd companies because the market of securities would collapse if the supply of shares extremely exceeded the demand. So far few enterprises have decided on transferring the whole company into a limited one. They rather separate one or some of their activities and estabilish an ltd for them.

3. *Transformation of a state enterprise* into a ltd company the shares of which would belong to the "Fund of state assets" or to hospitals, schools etc.

In the first case the Fund of state assets needs many well qualified experts who have experience in this field but they are not available in Hungary today.

In the second case state subsidies of hospitals, schools, etc. could be replaced by state shares and expenditure of these organisations could be financed from the dividends of these shares. The disadvantage of this solution is that on the one hand these organisations are interested in short term consumption and they would sell their shares, on the other hand they are not experts in investment management therefore they would probably be careless owners.

4. *Issue of "popular shares"*. The representatives of this idea state the following: public ownership is the ownership where the owner is the whole nation. Therefore - to speed up the privatisation - "popular shares" of the same nominal value should be

[2] The idea of devaluation of state assets also arose. In this case - for the purpose of accelerated privatisation - shares of public enterprises could be bought at a reduced price. This solution, however, would cause great inequalities.

issued and given to every Hungarian citizen. These securities could be spent on buying real shares of Hungarian companies.

Disadvantages of this concept are:

- A high proportion of the residents would sell the shares immediately and this would extremely reduce the price of the shares;
- For the previous reason the stock exchange couldn't work for a long time and this would hinder the prospering companies as well in getting capital from the stock exchange;
- The high volume of additional income at the residents (coming from selling shares and dividends) would cause further inflation.
- Issue of popular shares would reduce also the price of those shares which were bought at full price before the issue of popular shares.

For the time being a slow process of privatization is going on by the investment of foreigners, Hungarian private investors and residents. The way of accelerated liberalisation is under elaboration and the most likely solution will be a mixture of these alternatives. This will also be determined by the new parliament.

2. The Situation and Future of Hungarian Telecommunications

In this section some key figures for the Hungarian telecommunications system will be given as well as an overview over the current organisational background of provision of telecommunications services. Finally, the need for and the opportunities of liberalisation in the Hungarian telecommunications sector will be discussed.

2.1 General Information about the Hungarian Telecommunications System

In the following table some basic figures for the current status of telephony in Hungary are presented.

Table 1: Telephony indicators for Hungary

Telephone density	15.81 (per 100 inhabitants - 1988)	
Budapest:	37.46	
countryside:	10.40	
Main-line density	8.10	
Budapest:	20.35	
countryside:	5.05	
Direct dialing: (1988)	85.2 % of main lines 685 settlements out of the total 3.200 internationally: 107 countries	
Waiting list:	total	residential(1988)
Budapest:	241.045	223.659
countryside:	283.335	280.434
total	524.380	504.093
Average waiting time:	14 years	

The next table presents some basic information for the provision of telecommunications services other than telephony service.

Table 2: Indicators for other telecommunications services

Telex density	11.9 per 10 thousand inhabitants	
Data modems	3.4 per 10 thousand inhabitants	
Mobile service		dedicated networks only (in operation) pilot public service in Budapest
Telefax	2.500	1989; (opening Dec. 1st, 1987).
Burofax	39	1989; (opening Aug. 1st, 1985).
Teletex	10	1989; (opening Oct., 1987).
Videotex	3-400	1989; (opening Febr. 1st, 1989).
Minitex	ca. 100	1989; (opening Febr. 1st,1989).
Packet switching	ca. 50	1989; (opening Jan., 1989)
Telecommunications investments as a share of GDP (average between 1964-85)	3.11 thousandth	
Effectivity	30 main lines/employee (1984)	

2.2 Organisational Framework Today

The Hungarian Post and Telecommunications Company (HPTT) has a legal monopoly (ensured in the Post-Act of 1964) on postal, telecommunications services and on broadcasting activity. Besides, the Act declares that all the assets used in transmission of signals have to be in public ownership. Prior to January lst, 1989 it was the HPTT which had all the authority activity as well.

Beside the main services (post, telecom, broadcasting) HPTT runs secondary activities as well: construction, transportation, R&D, training, procurement, etc).

The total number of its employees amounts to 70 000, and 24 000 are working in the telecommunications branch.

HPTT has a three-tiered organisation: headquarters, 23 "directorates" (7 regional and 17 functional ones) and the executive tier.

The *future of Hungarian telecommunications* is determined by 3 components today:

- *The 10 year development program.* This government program intends to triple the main lines by the year of 2000 and will introduce nationwide direct dialing ;

- The new post- and telecom Act will come into force next year;

- The HPTT will be restructured in 1990.

2.3 Need for and Possibilities of Liberalisation in Hungary

The goal of liberalisation in the telecommunications policy of developed countries is to speed up introduction of new services, increase efficiency on the suppliers' side, improve the quality of the services, and decrease prices.

Liberalisation is less characteristic of the developing countries. Those developing (or in many cases newly industrialised) countries, which decided on speeding up development of telecommunications used a wide variety of economic means for this purpose: government subsidy, foreign loans, but privatisation as well.

What are the chances of liberalisation in the Hungarian telecommunications?

As we saw above, the Hungarian economy is undergoing a substantial liberalisational course today. The question is: how can this principle be implemented in telecommunications, namely:

- to what extent and in which parts of telecommunications can competition be introduced?
- to what extent must liberalisation be subordinated to the acceleration of development?

A monopoly-committee elaborated an Act-Proposal by the beginning of the year 1989 which has been discussed on several forums.

If we compare the tendencies of liberalisation in the developed countries and the Hungarian intentions similarities and differences can be found as can be seen from table 3.

How can the alternatives of the proposal be explained?

The separation of authority and operational activity was a decisive intention of the government. Prior to 1989 the HPTT belonged to the government directly but in 1989 (January lst) a new ministry was established (Ministry of Transport, Communications and Construction Industry) and HPTT was deprived from authority tasks which have been done by this ministry since then. In 1988 all the price-control activities were transferred from the HPTT to the National Price Office.

The monopoly-committee consisted of the employees of the HPTT and this fact left its mark on the proposal. As to the liberalisation of terminal equipments and maintenance of terminal equipments (two marginal areas) the committee easily submitted to liberalisational pressure.

In the case of telex-service and data transmission, however, the committee insisted on monopoly and the "yes" alternative is an outer recommendation.

In the case of network liberalisation the finance of development has high priority in the Hungarian thinking. The chance that liberalisation can help to speed up the development of telecommunications convinced the decision-makers: liberalisation can be an alternative for Hungarian telecommunications policy.

Table 3: Features of liberalization in developing countries and in Hungary

	developed countries	Hungary (proposal)
Separation of authority and operational function	yes	yes (on Jan 1, 1989)
Liberalisation of terminal equipments	yes	yes
Liberalisation of mainten-ance of terminal equipments	yes	yes
Liberalisation of data transmission	yes	yes, no alternatives
Liberalisation of telex service	differ	yes, no alternatives
Network liberalisation	yes no	A, B, C alternatives
Motivation of network-liberalisation	competition	finance
Direction of network liberalisation	alternative networks	territorial companies
Ensuring legal monopoly	no	yes, no alternatives

This aspect determined the direction of liberalisation: a regional liberalisation is intended to be implemented by which the national network would be split up into regional companies.

This solution has chances for the following reasons:

- A small company is more attractive for Hungarian companies which have small volume of spare capital to invest;

- Foreign investors still think Hungarian investments risky, therefore a small investment is less risky for them;

- Under present circumstances Hungarian decision-makers don't dare to make a big investment into one company (the performance of which was very poor so far); a "modul-scheme" seems to be more safe.

All in all the Act-Proposal contains the following alternatives:

A: All telephone services are provided under monopoly
(recommendation of trade union)

B: The local telephone network would be liberalised
(recommendation of monopoly-committee)

C: The whole domestic telephone service would be liberalised
(outside recommendation).

The regional liberalisation ensures regional (but not legal) monopoly which, I think, is better than legal monopoly but users benefit less from this than from competition of alternative networks. The idea of licensing of alternative networks rarely arose in Hungary. Why is this the case?

First there hasn't been a market economy and strong competition in Hungary yet, therefore the efficiency requirement has smaller significance. To give a second argument, let me cite Charles Pautrat (French PTT) who determined three stages of development of telecommunications from the aspect of demand:

1st stage: people want telephone

2nd stage: people want telephone of good quality

3rd stage: people want telephone of good quality at a good price.

By this classification Hungary is at the first stage, so what is most important today is to have access to the telephone network.

A second aspect for the three alternatives A, B, C mentioned above is whether there is

- a legal monopoly or
- a declaration of telecommunications state monopoly which would mean that telecommunications is a licensed activity and a government office has the right to licence suppliers.

The Act-Proposal recommends to abolish the paragraph of the old Act which declares that all the assets in telecommunications have to be in public ownership.

All in all the present version of the proposal contains a wide spectrum of alternatives which refers to the fact that recommendations of several lobbies of different interests got into the proposal and the standpoint of the government (including the new ministry) is under formation.

The Act-Proposal was to be submitted to the Parlament December 1989 and come into force on 1st of January 1990. The opposition parties - who don't accept the legality of the present parliament - however, slowed down the legislation procedure. Therefore the Post-Act will be discussed by the new parliament, after the free elections. In order to enable the investors to invest in telecommunications as soon as possible the Act will probably be modified to ensure a legal base for that.

3. Restructuring of the Hungarian PTT

A year ago the government decided on restructuring of the PTT the main goal of which is the separation of postal and telecom services. The reason for separation can be explained as follows:

a) *Cross-subsidization* of the postal services from telecommunications hinders the development of telecommunications (the yearly sum of the subsidy from telecom to post equals the subsidy from the state to telecom).

b) Raising of *loans* needs *a clear cost-distribution.* Till the middle of the 1980's the PTT financed its investments from internal sources and a state subsidy and rarely applied for credits. In the last years, however, finance by credits became more frequent. Financial separation of post and telecom services is a precondition for the next Wold Bank credit as well.

c) The probable liberalisation of telecom also makes is necessary that the company must be aware of the real unit costs. Therefore separation of the services is a preparation for the new market situation.

The separation of the two main activities forced the PTT to think over the restructuring of the whole PTT and the following recommendation was made by the PTT (the final decision will be made by the government): Three (traditional) state enterprises would be established for the basic services for the time being. These are

- The Post enterprise,

- The Telecom enterprise, and

- The Broadcasting enterprise.

Later on the broadcasting and telecom company, respectively, would be transformed into incorporated companies. The Restructuring Committee regards this transformation as urgent as to the Telecom enterprise (and it would be carried out next year) in order to initiate external investments in telecommunications.

For the competitive area of the PBX services (installation, maintenance, etc) the Committee recommends to set up a company in Budapest at the beginning of next year.

In order to get more flexible companies the Committee recommends the separation of the secondary activites as well. Most of these activities are run in different directorates today but restructuring would mean splitting up some directorates as well. The new companies would be joint ventures of the Post and Telecom companies or they would belong to one of them; it depends on the fact to what extent the activity is used by one company or by both. The following activities would be involved in the separation: network-installation, building construction, manufacturing, technical design, R&D and transportation.

Appendix: Comparison of Joint Ventures in the East European Countries

Countries	Establishment of the first joint venture (year)	No. of joint ventures (today)	Authorising organisation	Tax rate (% on profit)	Tax discount	Tax on tax (%)	Repatriation of the profit
Bulgaria	1980	20	If foreign ownership exceeds 50%: government. Other-wise: registration only	30	J.v. can apply for being tax free for the first 5 years	10-15	-
Chekoslo-vakia	1985	19	ministries	40	-	-	-
Yugoslavia	1968	230	It varies republic by republic	10	-	-	-
Poland	1976	53	Agency of Foreign Investment	40	First 3 years are tax free, j.v. can apply for it for the next 3 years, tax base can be reduced by reinvestment	30	85% of export surplus
Romania	1971	5	Government	30	1. year: tax free 2. year: 15 % - 20 % reduction if profit is not repatriated	10	-
Soviet Union	1988	270	Government	30	First two years: tax free Far East: first 3 years tax free, further reductions later	20 (reduction maybe)	profit of export in hard currency
Hungary	1977	310	If foreign ownership exceeds 50%: Ministry of Trade and Min. of Finance; otherwise: registration only.	up to 3 million HUF: 40% above 3 mil.: 50%	If foreign ownership exceeds 20%, 20% tax reduction for the first 5 years. In specific activities tax-reduction can be 100% for the first 5 years, from 6. year: 60 % reduction	individuals: personal income tax companies: -	full repatriation is ensured

East German and Albania don't stimulate establishment of joint ventures.
Source: Heti Világgazdaság, Budapest, june 3, 1989

10

Regulatory Instruments and EEC Powers to Regulate Telecommunications Services in Europe

Joachim Scherer

Baker & McKenzie
and Faculty of Law, Frankfurt University
Frankfurt

European telecommunications *policy* proposals are gradually being transformed into European telecommunications *law*.

In its Green Paper on the Development of the Common Market for Telecommunications Services and Equipment - "the Green Paper" -[1], and in a follow-up Communication[2], the Commission of the European Communities (- "the Commission" -) has announced that it will implement some of its main telecommunications policy proposals by using the Community law instrument of "directives".

---

1 COM 87 (290) final, 30 June 1987

2 COM 88 (48) final, 9 February 1988: Towards a Competitve Community-Wide Telecommunications Market in 1992: Implementing the Green Paper on the Development of the Common Market for Telecommunications Services and Equipment.

Up to now, the Commission has adopted the

> "Commission Directive on Competition in the Markets in Telecommunications Terminal Equipment" - "Terminal Equipment Directive" -[3]

and has submitted several proposals for directives and draft directives with regard to the provision of telecommunications services, terminal equipment and public procurement in the telecommunications field[4]. It appears, from the Commission's announcements and these proposals, that the Terminal Equipment Directive is merely the beginning of a series of regulatory measures with the objective to create a Common Market for Telecommunications Services and Equipment by 1992.

This paper explores, in a first step, the various types of regulatory instruments which the Community has at its dis-posal in the telecommunications field, in particular directives, decisions and technical standards (part 1).

Secondly, the respective powers of the Commission and of the Council to issue directives in the telecommunications field will be delineated (part 2). Part 3 describes briefly the Commission's "new approach" to using directives and technical standards as instruments of harmonization.

In conclusion, the paper attempts to assess the legal and political premises of the evolving European telecommunications policy and their consequences for the policy makers (part 4).

1. Regulatory Instruments

1.1 Overview

The form and manner in which the Community institutions may exercise their respective regulatory powers (not only) in the telecommunications field are governed, inter alia, by Art. 189 (1) EEC Treaty.

3 O.J. L 131, 27 May 1988

4 See, inter alia, Revised Proposal for a Council Directive on the establishment of the internal market for telecommunications services through the implementation of Open Network Provision, COM (89) 325 final, 10 August 1989; Draft Commission Directive on Competition in the Markets for Telecommunications Services, 28 June 1989; Amended Proposal for a Council Directive on the procurement of entities operating in the water, energy, transport and telecommunications sectors, COM (89) 380 final - SYN 153 -, 31 August 1989; Proposal for a Council directive on the approximation of the laws of the Member States concerning telecommunications terminal equipment, including the mutual recognition of their conformity, COM (89) 289, 27 July 1989.

Art. 189 (1) EEC Treaty provides that

> "in order to carry out their task the Council and the Commission shall, in accordance with the provisions of this Treaty, make regulations, issue directives, take decisions, make recommendations or deliver opinions".

The four types of Community acts - regulations, directives, decisions and recommendations/opinions - are distinguishable according to their binding force and the scope of their applicability.

A *regulation* is binding in its entirety and it is directly applicable in all Member States. It is characterized by its general application which means that it may be applicable to an undetermined number of cases.

A *decision* is binding in its entirety but, unlike a regulation, it is only binding upon those to whom it is addressed.

Recommendations and *opinions* differ from both regulations and decisions with respect to their binding force: Neither recommendations nor opinions have binding force.

Directives differ from regulations with respect to the scope of their binding force. According to Art. 189 (3) EEC Treaty, a directive is merely binding

> "as to the result to be achieved, upon each Member State to which it is addressed, but shall leave to the national authorities the choice of form and methods".

According to the Commission's Green Paper and the Commission's regulatory and preparatory activities to date, European telecommunications policy will essentially be implemented by means of Community law directives. Therefore, the legal scope and effects of this regulatory instrument will now be described in more detail:

1.2 Directives and their Implementation

Whereas regulations are directly applicable and have the force of law in all Member States without the need of transformation or confirmation by national authorities, directives require implementation on the national levels. In practice, directives have become the main instrument for the approximation of national laws: Unlike regulations, directives are not meant to be an instrument of *uniformity* but rather an instrument for the creation of *equivalence* of national laws.

Directives are addressed to and binding upon the Member States and require the adaptation of national laws by "the national authorities". The term "national authorities" encompasses all three branches of national government.

The wording of Art. 189(3) EEC-Treaty is not very clear: The distinction between the "result" to be achieved by a directive on the one hand and the "form" and "methods" of the implementation by the Member States on the other hand is untenable both from atheoretical and from a practical viewpoint. "Results" and "methods" to achieve them can hardly be distinguished: Any result is at the same time a method to achieve another result and any method applied in order to achieve a result is at the same time a result .[5] Furthermore, no clear boundary line can be drawn between "form" and "methods" of the implementation of directives.[6]

Although the wording of Art. 189(3) EEC Treaty seems to suggest that the regulatory scope of directives is limited to the mere definition of "results" (in German: "Ziel", in French: "résultat", in Italien: "risultato", in Dutch: "resultaat") to be achieved, both Community practice and legal doctrine have long acknowledged that no such limitation exists: The Community organs may regulate a given field in detail by directives if such regulation is necessary in order to achieve a goal under the EEC Treaty.[7]

Consequently, the Community organs are free to decide about the regulatory scope of their directives, subject only to the restrictions contained in the Treaty.

These restrictions are, in principle, twofold. They result from

- the various Treaty provisions which empower the Council or the Commission to issue directives (p.e. Art. 100 or Art. 90(3) EEC Treaty) and

- the principle of proportionality.

The Treaty provisions empowering the Community organs to issue directives define the competence of the respective organs to issue directives and the permissible goals of such directives. The principle of proportionality requires that the directives issued in pursuit of these goals must be appropriate and necessary (i.e. the least restrictive means) in order to achieve the respective policy goals defined by the Treaty.[8]

---

5 Cf. Dieter Oldekop, Die Richtlinien der EWG, 1968, p. 140. See also Hans-Peter Ipsen, Richtlinien-Ergebnisse, in: Festschrift Ophüls, 1965, S. 67, 71 ff.

6 Cf. Case 188-190/90, France et al. v Commission, (1982), ECR 2545, for an analysis of this decision see below, part 2.4.

7 For an analysis of the interdependence between the competence rules of the Treaty and the Community's power to issue directives see Manfred Zuleeg, Das Recht der Europäischen Gemeinschaften im innerstaatlichen Bereich, KSE Vol. 9, 1969, pp. 261-294.

8 For attempts in legal doctrine to restrict the Community's powers to issue detailed, substantive directives see Oldekop, loc. cit., pp. 144 seq; see also Zuleeg, loc. cit., pp. 270 seq.

The permissible scope of Community law directives appears to be a function of the regulatory area in question: If and when the Treaty clearly defines regulatory goals (p.e. Art. 54 III g), directives may contain very detailed measures in order to achieve these goals. If and when the Treaty sets only very general and abstract goals (p.e. Art. 100),the permissible regulatory scope of directives would seem to be slightly reduced[9], although in practice, directives may well be sufficiently detailed to make the transformation a mere formality.

According to the European Court of Justice, the decisions concerning

- the respective national authority in charge of the implementation of a directive,
- the implementation procedure and
- the legal quality of the national measure implementing the directive

are left to the Member States.[10]

Whether or not a Member State implements a Community law directive by means of a national statute, a government ordinance, or an administrative rule, is a matter of "choice" by the national authorities.[11] This freedom of choice is, however, subject to the general provisions of the Treaty: Under Art. 5 EEC Treaty, the Member States are obliged to take *appropriate* measures to ensure the fulfilment of the obligations arising out of the Treaty. The European Court of Justice has emphasized the relevance of this duty in its *Royer* decision which stated that the Member States are obliged to choose within the freedom of choice established by Art. 189 (3) EEC Treaty the form and methods which are best suited to ensure the "effet utile" of directives in view of their respective goals.[12]

In spite of the freedom of choice awarded to the Member States under Art. 189 (3), implementation of a directive must be based upon a *legally binding act*, i.e. the setting of new legal rules or the amendment of existing legal rules. A mere change of factual administrative practices without any legal underpinning does not suffice.[13]

The Member States are obliged to implement Community law directives within the time limits imposed by the directives. Failure on the part of a Member State to implement a

9 Cf. Zuleeg, loc. cit., pp. 261 seq.
10 Case 96/81, Commission v. Netherlands (1982), ECR 1791 (1804).
11 For a discussion of the problem wheter or not Community law directives may be implemented by mere technical standards which are only internally binding upon administrative authorities, see Ulrich Beyerlin, Umsetzung von EG-Richtlinien durch Verwaltungsvorschriften? in: EuR 1987, pp. 126-145
12 Case 48/75, Royer, (1976), ECR 497.
13 Case 300/81, Commission v. Italy, (1983), ECR 449.

directive within the time limit imposed constitutes a violation of the EEC Treaty.[14]

After implementation, the Member State is no longer free to amend the statutes, rules and/or regulations implementing the directive. This blocking effect of Community law directives is a consequence of the primacy of Community law over national laws.[15]

Failure of a Member State to implement the directive as a whole or in part constitutes a violation of Treaty obligations and will lead to the institution of an infringement proceeding under Art. 169 EEC Treaty.

The infringement proceeding is a three-step procedure with the objective to ensure the fulfilment of the Treaty obligations by the Member States.

The *first step* consists of a warning letter from the Commission to the Member State giving the State the opportunity to submit its observations concerning the alleged violation of the Treaty. If the Commission considers that the Member State has failed to fulfil an obligation under the Treaty, it shall, as a *second procedural step*, deliver a reasoned opinion on the matter. If the State concerned does not comply with the opinion within the period laid down by the Commission, the latter may, as a *third step*, bring the matter before the Court of Justice.[16]

The Member States cannot invoke national regulations, customs or other circumstances of their national legal systems in order to justify non-implementation or belated implementation.[17]

Furthermore, non-implementation may neither be justified on the grounds that institutional reforms within the Member State prevented timely implementation[18] nor on the grounds that other Member States have failed to implement the directive.[19]

---

14 Cf. Case 79/72, Commission v. Italy, (1973), ECR 667; Case 52/75, Commission v. Italy, (1976), ECR 227.

15 Df. Eberhard Grabitz, in: Kommentar zum EWG-Vertrag, Art. 189, Note 57.

16 For an analysis of the procedure under Art. 169 EEC-Treaty see Beate Ortlepp, Das Vertragsverletzungsverfahren als Instrument zur Sicherung der Legalität im europäischen Gemeinschaftsrecht, 1987; see also Claus-Dieter Ehlermann, Die Verfolgung von Vertragsverletzungen der Mitgliedstaaten durch die Kommission, in: Festschrift für Hans Kutscher, p. 135 seq. and A.C. Evans, The Enforcement Procedure of Art. 169 EEC, in: European Law Review 1979, p. 442.

17 Case 52/75, Commission v. Italy, (1976), ECR 277; see also: Case 100/81, Commission v. Netherlands, (1982), ECR 1937; Case 307/81, Commission v. Belgium, ECR 1983,467.

18 Case 68-72/81, 83/81, Commission v. Belgium, (1982), ECR 153, 163, 169, 175, 183.

19 52/75, Commission v. Italy, (1976), ECR 277.

1.3 Directives and Their "Direct Effect"

While the direct effect of *regulations* poses no problems, the direct effect of *directives* has been subject to considerable legal controversy[20] and has met with resistance from some national courts.[21] The direct application of a Community regulation means, according to the European Court of Justice, "that its entry into force and its application in favour of or against those subject to it are independent of any measure of reception into national law".[22] This direct application of regulations is functionally related to their direct effect: A directly effective provision is one "producing direct effects and creating individual rights which national courts must protect".[23]

Regulations are characterized by an inherent direct effect:[24] According to the Court, "regulations are directly applicable and therefore by virtue of their nature capable of producing direct effects".[25]

Although Art. 189 (3) EEC Treaty appears to indicate that directives are not directly applicable and should therefore not have direct effects, the case law of the Court is unequivocal in confirming the direct effect of directives.According to the Court, Art. 189 EEC Treaty does not constitute an obstacle to a direct effect of a provision of a directive or a decision because this Treaty provision concerns merely the *manner* in which the Community powers may be exercised but not their effects. In the *Grad* case, the Court held:

> "... Although it is true that by virtue of Article 189, regulations are directly applicable and therefore by virtue of their nature capable of producing direct effects, it does not follow from this that other categories of legal measures mentioned in that article can never produce similar effects".[26]

20 Cf. A.J. Easson, Can directives impose obligations on individuals? in: ELR 1979, 67; E.-W. Fuß, Die Richtlinie des Europäischen Gemeinschaftsrechts, DVBl 1965, p. 378 seq.; Rambow, Probleme bei der Durchführung von Richtlinien der EWG, DVBl 1968, p. 445; B. Boulois, L'applicabilité directe des directives, RMC 1979, 104; M. Zuleeg, Die Rechtswirkung europäischer Richtlinien, ZGR 1980, 467 (474–477); Seidel, Direktwirkung von Richtlinien, 1983; id., Die Direkt- oder Drittwirkung von Richtlinien des Gemeinschaftsrechts, NJW 1985, p. 517 seq.; for a comprehensive analysis see Andreas Oldenburg, Die unmittelbare Wirkung von EG-Richtlinien im innerstaatlichen Bereich, 1984.

21 Cf. Conseil d'Etat, Decision dated 22 December 1978 (Cohn-Bendit Case), Rev. trim. dr. europ. 1977, 822 with annotations by Bieber, in: EuR 1979, 292, Petzold, in: DVBI 1980, 126; see also Tomuschat, EuGRZ 1979, 257 (258 seq.). See also Bundesfinanzhof, BFHE 133, 470.

22 Case 50/76, Amsterdam Bulb, (1977), ECR 137, 146; see also 94/77 Zerbone, (1978), ECR 99.

23 Case 26/62, van Gend en Loos, (1963), ECR 1, 13; see also A. Dashwood, The Principle of Direct Effect in European Community Law, 16 Journal of Common Market Studies 229 (1978).

24 Cf. Case 50/76, Amsterdam jBulb, (1977), ECR 137, 146: see also Case 31/78, Bussone, (1978), ECR 2429, 2444: "The direct effect (is) inherent in regulations and other rules of Community law".

25 Case 9/70, Grad, (1970), ECR 825, 837.

26 Case 9/70, Grad, (1979), ECR 825, 837.

There are, however, certain differences between the direct effects of regulations and directives: The provisions of regulations are *per se* directly effective. By contrast, a directive is viewed by the Court as "a legislative act which has no automatic effect in its entirety".[27] Rather, the provisions of directives are directly effective only if and when certain criteria established by the Court's case law are met: In particular, a direct effect can only result from a directive provision which is unconditional and sufficiently clear and precise to be capable of producing a direct effect.[28]

2. Regulatory Powers in the Telecommunications Field

The Community's power to regulate telecommunications to the extent necessary for the completion of the Internal Market appears not to be controversial in principle.

Notwithstanding this apparent consensus, however, the regulatory powers of the Council on the one hand and the Commission on the other hand are highly controversial.

Directives can be issued by the Commission or by the Council depending upon their legal basis in the EEC Treaty.In the telecommunications field, Art. 100 a EEC Treaty may serve as a basis for Council directive whereas Art. 90 (3) EEC Treaty could arguably be used as a basis for Commission directives.

The Commission directive of 16 May 1988 on competition in the markets in telecommunications terminal equipment[29] is based on Art. 90 (3) EEC Treaty. Other Community law directives, including the directive for Open Network Provision, may be based on Art. 100 a EEC Treaty.The choice of the legal basis for directives in the tele-communications field is not a mere technicality. It may well prove to be crucial for the speedy implementation of the European telecommunications policy as enunciated in the Green Paper: The enactment of Commission directives is in general less time-consuming than the enactment of a Council directive, in particular in view of the often conflicting interests of the Member States in the telecommunications field.

The following sub-chapters will describe and delineate the scope of Art. 90 (3) and 100 a EEC Treaty as the legal basis for directives in the telecommunications field.

27 Case 41/74, von Duyn, (1974), ECR 1337, 1348.

28 Cf. Case 9/70, Grad, (1979), ECR 825, 838 with respect to the direct effect of a decision. The same applies to directives, cf. Case 8/81, Becker, (1982), ECR 53, 70 seq; see also Case 126/82, Smit Transport, (1983), ECR 73, 88; Case 271/82, Auer, (1983), ECR 2727, 2744. See also Gerhard Bebr, Development of Judicial Control of the European Communities, 1981, p. 584-588.

29 O.J. L 131, 25 May 1988.

2.1. Objectives of Art. 90 (3) and Art. 100, 100 a EEC Treaty

Art. 90 (3) EEC Treaty is part of the Treaty's "Rules on Competition" and provides that

> "the Commission shall ensure the application of the provisions of this Article and shall, where necessary, address appropriate directives or decisions to Member States".

Art. 100 a EEC Treaty is an amendment to Art. 100 EEC Treaty concerning the approximation of laws through Council directives. Under Art. 100 EEC Treaty, the Council shall, acting ***unanimously*** on a proposal from the Commission, issue directives for the approximation of such provisions of national laws "as directly affect the establishment or functioning of the common market".The Single European Act has amended Art. 100 by Art. 100 a EEC Treaty which has been termed one of the most important of the Act's provisions aimed at establishing an "internal market" by 31 December 1992.[30]

2.2. Approximation of Laws under Art. 100, 100 a EEC Treaty

2.2.1. Terminology

The EEC Treaty uses the terms "approximation" (p. e. in Art. 100, 100 a), "coordination" (p. e. in Art. 54 (3) g, 56 (2)) and "harmonization" (p. e. Art. 99).

In spite of numerous attempts in legal doctrine to differentiate between these three categories, no clear boundary lines appear to exist.

It is now generally acknowledged that the three terms are to be considered as synonyms and designate the same task of the EEC.[31]

This task and the common denominator of "approximation", "coordination" and "harmonization" is to abolish all barriers within the Common Market which result from the existence of diverging legal and administrative rules in the various Member States.

This harmonization (= approximation = coordination) is achieved by two types of measures which may be distinguished from a theoretical point of view but which, in practice, are inextricably intertwined:

(1) the abolishment of diverging national legal and administrative rules and

30 Cf. C.D. Ehlermann, The Internal Market Following the Single European Act, in: CMLRev 1987, p. 361 (381).

31 Cf. Hans Claudius Ficker, Zur Angleichung der Gesellschaftsrechte in den erweiterten europäischen Gemeinschaften, in: Law and International Trade - Festschrift für Clive M. Schmitthoff, Frankfurt 1973, p. 157 (161); Berthold Goldmann, Le Rapprochement des Legislations, in: Les Novelles - Droit des Communautés Européennes, Brussels 1969, p. 883.

(2) the replacement of such rules by new Community rules which are set up in view of the objective to create a unified single market.

2.2.2 Scope of Art. 100 EEC Treaty

The possible scope of measures concerning the approximation of provisions laid down by law, regulation or administrative action in Member States results from the function of Art. 100, 100 a EEC Treaty.

Approximation of laws is required in order to establish the Common Market and ensure its operation ("functioning" cf. Art. 100 EEC Treaty).

Consequently, the scope of permissible approximation of laws is implicitely defined by the goals established in Art. 2 EEC Treaty. In order to achieve these goals, it does not suffice to eliminate customs duties and quantitative restrictions on the import and export of goods and of all other measures having equivalent effect between Member States and to establish a common customs tariff and a common commercial policy toward third countries. These measures must be complemented by the abolition of obstacles to freedom of movement for persons, services and capital (cf. Art. 3 c EEC Treaty).

The specific role of approximation of laws in this context results from the fact that the majority of national laws and regulations were enacted in view of national markets taking into account national public interests which are not necessarily congruent with the public interest of the Community. National laws reflect national policies and are based on national interests.

The task of approximation of laws was and is to replace this national legal framework by rules of Community law which are based on Community policy goals as set forth in the Treaty and which reflect the Community's public interest as determined by the Community organs according to the decision making procedures under the Treaty.[32]

Apart from this coordination function, the approximation of laws guarantees the functioning of the Common Market.[33]

2.2.3 Prerequisites for Approximation of Laws under Art. 100 EEC Treaty

Art. 100 EEC Treaty establishes three substantive prequisites for the approximation of laws: The Council shall issue directives for the approximation of

32 Cf. Hans Claudius Taschner, Art. 100, Note 3, in: Groeben/Boeckh/Thiesing/Ehlermann, Kommentar zum EWG-Vertrag, 3rd edition.

33 Case 33/76, Rewe v. Landwirtschaftskammer Saarland, (1976), ECR 1989, Ground 5.

(1) provisions laid down by law, regulation or administrative action in Member States

which

(2) directly affect

(3) the establishment or functioning of the Common Market.

at 1) The objects of approximation of laws are "provisions laid down by law, regulation or administrative action in Member States". Provisions of law are all legal rules which have been formally adopted by the competent national authorities, including rules and regulations adopted by the legislative and the executive branches of government as well as rules adopted by the case law of the respective national courts. Furthermore, the approximation of laws includes the harmonization of technical standards in various fields of industry and technology. This harmonization has become the object of the Community's "new approach" as defined, inter alia, in the White Paper.[34]

In sum, the terms "provisions laid down by law, regulation or administrative actions" are construed very broadly.

It is generally acknowledged that the approximation of laws by a Council directive does *not* require the existence of national laws in every Member State.

The functioning of the Common Market may already be affected by the existence of national laws in only just one single Member State.[35]

at 2) Art. 100 EEC Treaty requires that the existing national law(s) must *directly affect* the establishment or functioning of the Common Market. Given the width of the term "Common Market", the scope of the approximation of laws is to a large extent determined by the development of criteria concerning the "direct" effect of national laws upon the establishment and the functioning of the Common Market.

The European Court of Justice does not require a clear and present negative effect upon the Common Market. Rather, the Court is satisfied if and when national laws are *likely to affect* the functioning of the Common Market.[36]

34 See below, sub 3.
35 Cf. Langeheine, Art. 100, Note 20, in: Grabitz (ed.), Kommentar zum EWG-Vertrag.
36 Case 33/76, Rewe v. Landwirtschaftskammer Saarland (1976), ECR 1989, Ground 5.

In assessing this likelihood, the Commission and the Council would seem to have a considerable political discretion which may only to a limited extent be scrutinized by the Court.[37]

at 3) The term "Common Market" is not defined in the Treaty of Rome. Its main features are, however, sketched out in Arts. 2 and 3 EEC Treaty (see above).

The European Court of Justice has indicated that it views the concept of the "Common Market" as mentioned in Art. 100 in the light of the preamble and the general objectives of the Treaty.[38]

2.2.4 Scope of Art. 100 a EEC Treaty

Art. 100 EEC Treaty has been amended and, in part, supplemented by Art. 100 a EEC Treaty which is a *lex specialis* to Art. 100 EEC Treaty. It enables the Council to enact measures for the approximation of national laws on the basis of a qualified majority rather than subject to the unanimity requirement of Art. 100 EEC Treaty. In essence, Art. 100 a EEC Treaty contains special procedural and substantive rules with regard to those approximation measures which aim at the achievement of the objectives set out in Art. 8 a.

The main objective of Art. 100 a is the dynamisation and acceleration of the approximation of laws. This objective will have to determine the interpretation of Art. 100 a EEC Treaty.

Given the controversies surrounding the adoption of Art. 100 a as reflected by several declarations in conjunction with the Single European Act and the resulting ambiguities in the wording of Art. 100 a EEC Treaty, the precise delineation of the scope of this provision will become crucial.

2.3 Commission Directives on the Basis of Art. 90 (3) EEC Treaty

In order to delineate the respective regulatory powers of the Commission and the Council under Art. 90 (3) and Arts. 100, 100 a EEC Treaty, it may be recalled that the European Court of Justice has laid down several basic principles for the exercise of legislative powers under Art. 90 (3) EEC Treaty and their relation to other Treaty provisions.

---

37 Cf. Leleux, Cah. Dr. Europ. 1968, p. 142.

38 Case 43/75, Defrenne II (1976) ECR 455, Ground 61/64; see also Case 91/81, Commission v. Italy (1982) ECR 2133, Ground 11

In the Transparency Directive Case[39], the Court has held that Art. 90 (3) EEC Treaty does not restrict the legislative powers of the Commission to the adoption of directives which merely deal with a specific situation in one or more Member States. The Court rejected the submission of plaintiffs that Commission directives were not of the same nature as those adopted by the Council. The Court found that the limits of the powers conferred on the Commission are to be inferred not from a general principle,

> "but from an interpretation of the particular wording of the provision in question, in this case Article 90 analysed in the light of its purpose and its place in the scheme of the Treaty".[40]

Consequently, the Commission was entitled to issue a directive of general application to all Member States with the objective to liberalise the telecommunications services market.

A second basic principle for the application of Art. 90 (3) EEC Treaty follows from the Court's holding that the existence of concurrent legislative powers of the Council does not exclude the adoption of a directive of general application based on Art. 90 (3) EEC Treaty.

As the Court has pointed out, Art. 90 has a specific objective which justifies the adoption of Commission directives under Art. 90 (3) EEC Treaty in spite of concurring legislative powers of the Council:

> "Article 90 concerns only undertakings for whose actions States must take special responsibility by reason of the influence which they may exert over such actions. It emphasizes that such undertakings are subject to all the rules laid down in the Treaty, subject to the provisions contained in paragraph 2; it requires the Member States to respect those rules in their relations with those undertakings and in that regard imposes on the Commission a duty of surveillance which may, where necessary, be performed by the adoption of directives and decisions addressed to Member States".[41]

The third basic principle for the exercise of legislative powers under Art. 90 (3) EEC Treaty by the Commission follows from the objective of Art. 90 (3). According to the Court, the power conferred on the Commission by this provision is

> "limited to the directives and decisions which *are necessary to perform effectively the duty of surveillance* imposed upon it by that paragraph".[42]

---

39 Case 188 - 190/80, French Republic et al. v. Commission, ECR (1982) 2545.
40 Case 188 - 190/80, (1982), ECR 2545 Ground 6.
41 Case 188 - 190/80, (1982), ECR 2545, Ground 12.
42 Case 188 - 190/80, ECR 2545, Ground 13 (emphasis added).

The EEC Treaty does not *oblige* the Commission to adopt a directive under Art. 90 (3) EEC Treaty with regard to the liberalisation of telecommunications services. Rather, the Treaty leaves to the Commission a considerable amount of political discretion in deciding whether or not it prefers to exercise its supervisory function by means of a directive. In particular, the Commission has a considerable scope of discretion with regard to information requirements imposed upon the Member States by directives issued under Art. 90 (3) EEC Treaty.[43]

3. Directives and Technical Standards

Notwithstanding the competence of both Commission and Council to issue Community law directives in the telecommunications field, telecommunications law is, to a large extent, implemented and complemented by technical standards.

Telecommunications law and standards are inextricably intertwined.

Harmonization of technical standards is necessary in order to achieve compatibility, protection of health and safety, data protection, and in order to remove national standards which are aimed at creating trade barriers or which have the side-effect of creating trade barriers.

The harmonization of telecommunications standards is subject to the Commission's "new approach" to standardization. This new approach, which has been defined by a Council Resolution on the basis of a Commission Proposal[44], was endorsed by the Commission's White Paper on the completion of the Internal Market and has been applied in the field of telecommunications Law.

Essentially, the new approach stipulates that the Council should limit itself to define the essential objectives and requirements in the general interest by means of directives and should off-load technical and executive matters.

The task of defining technical specifications is entrusted, under the new approach, to European standardization bodies. Executive functions, such as the monitoring of both European and national standards and the adaptation of existing directives to technical progress, are increasingly entrusted to the Commission.

In the telecommunications field, the "new approach" has been taken, inter alia,

43 Case 188 - 190/80, (1982), ECR 2545, Ground 18.
44 Council Resolution of May 7, 1985 on an new approach to technical harmonization and standards, OJ 1985, C 136/1.

- in the Council Directive of July 24, 1986 on the initial stage of the mutual recognition of type approval for telecommunications equipment[45] and

- in the Commission's proposal for a Council Directive on the establishment of the internal market for telecommunications services through the implementation of Open Network Provision (ONP).[46]

The Council Directive on the initial stage of the mutual recognition of type approval for telecommunications terminal equipment (86/331/EEC) provides for the mutual recognition of conformity test results concerning mass-produced telecommunications terminal equipment if such equipment is in accordance with common conformity specifications.

These comformity specifications must comply with certain "essential requirements" wich are set forth in the directive. The essential requirements are

- user safety,
- safety of employees of public telecommunications network operators,
- protection of public telecommunications networks from harm and
- interworking of terminal equipment, in justified cases

The "common conformity specifications" are drawn up by the European Conference of Postal and Telecommunications Administrations (CEPT) in cooperation with specialized European standardization organizations such as the European Committee for standarization (CEN), the European Committee for Electrotechnical Standardization (CENELEC) and the European Telecommunications Standard Institut (ETSI).

The proposal for a Council directive on "Open Network Provision" (- ONP -) sets forth, inter alia, a procedural framework for the definition of ONP conditions. The objective of ONP conditions is to facilitate the provision of telecommunications services by private operators using public telecommunications networks and/or public telecommunications services.

According to the proposed directive, ONP conditions must comply with certain basic principles, defined in the directive, and must not restrict access to public telecommunications networks or public telecommunications services except for reasons based on certain "essential requirements" (e. g. security of network operations, maintenance of network integrity).

---

45 OJ L 217/21.
46 COM (89) 325 final - SYN 187, August 19, 1989, - not yet published in the Official Journal -.

Within the limits of these "basic principles" and "essential requirements", the ONP conditions will be defined, according to the proposed directive, in a multi-stage procedure, during which the ETSI will draw up European standards upon the Commission's request.

In order to obtain legally binding force, however, these standards must be accepted by the Commission and published in the Official Journal of the European Communities. Only then will the Member States be obliged to consider those standards as suitable for Open Network Provision conditions and will be required to presume that a service provider who complies with those standards fulfills the relevant essential requirements.

Both the Council Directive on the initial stage of the mutual recognition of type approval for telecommunications equipment (86/361/EEC) and the proposed ONP directive are characterized by a legislative reference technique whereby the legislative acts do not define technical standards in detail but rather set forth general principles and make reference to technical standards to be defined by private standardization bodies.

Whereas Directive 86/361/EEC appears to foreclose Commission scrutiny of common conformity specifications (cf. Art. 6 [1]), the proposal for an ONP directive empowers the Commission to take the ultimate decision by which the "private" standards established by the European standardization organizations are granted legally binding effect.

In view of both the rule of law and the principle of "competence attribué", the adoption of legally binding rules by private standardization organizations without an explicit decision by a Community organ would seem to be inadmissible under Community law.[47]

4. European Policy-Making in the Telecommunications Field: Preliminary Remarks

European policy-making in the telecommunications field is predicated upon specific legal (4.1) and political (4.2) factors which must be taken into account when assessing the benefits and, in particular, the (political) costs of the European telecommunications policy (4.3).

47 The legal problems of the reference technique are being discussed in the Federal Republic of Germany with regard to the principles of democracy and the rule of law, cf. Wolf-Rüdiger Schenke, Die verfassungsrechtliche Problematik dynamischer Verweisungen, in: NJW 1980, p. 743 - 749; Peter Marburger, Die gleitende Verweisung aus der Sicht der Wissenschaft, in: Verweisung auf technische Normen in Rechtsvorschriften, Berlin 1982, p. 27; see also BVerfG NJW 1984, p. 1225.

4.1 Legal Premises

One of the crucial legal factors to be included in any political cost-benefit analysis of European telecommunications policy is the principle of the supremacy of Community Law.

This supremacy principle is well established in the jurisprudence of the Court and was emphasized in Costa v. ENEL:

> "By contrast with ordinary international treaties, the EEC Treaty has created its own legal system which, on the entry into force of the Treaty, became an integral part of the legal systems of the Member States and which their Courts are bound to apply. By creating a Community of unlimited duration, having ... powers stemming from a limitation of sovereigty, or a transfer of powers from the States to the Community, the Member States have limited their sovereign rights, albeit within limited fields, and have thus created a body of law which binds both their nationals and themselves".[48]

The second legal factor to be taken into account in assessing European telecommunications policy-making is the implementation requirement: To the extent that the European telecommunications law is contained in community law Directives, it requires implementation on the national level.

The third legal factor shaping the European telecommunications policy and determining its political costs for the Member States, is the legal obligation, resulting from Art. 8 a EEC Treaty to establish, by 31 December 1992, the "Internal Market", namely

> "an area without internal frontiers in which the free movement of goods, persons, services and capital is ensured in accordance with the provisions of this Treaty".

4.2 Political Factors

Among the multitude of political factors which shape, spur and sometimes delay or prevent European telecommunications policy-making, the following appear to be of particular relevance:

First, the "liberalization incentive" which is a consequence of both the EEC Treaty rules on the free provision of (telecommunications) services as well as the competition rules of the Treaty and of the internal market objective.

From its early beginnings in the Green Paper, European telecommunications policy was not based on political compromises on the lowest common denominator of national tele-

48 (1964) ECR 585 (598).

communications policies. Rather, European telecommunications policy was and still is aimed at the removal of traditional exclusive rights of the Telecommunications Administrations. The removal of these exclusive rights and their replacement by (regulated) competition between Telecommunications Administrations and private providers of telecommunications networks, services and terminal equipment is, in its economic effect, a "liberalization" of a hitherto monopolized market.

This liberalization incentive is being counterbalanced, to varying degrees, by the reluctance of some, if not all of the Member States to divest themselves of their traditional regulatory powers in the telecommunications field. The high degree of the Member States' political sensitivity vis-à-vis the conceivable role of the EEC Commission as a telecommunications regulator is reflected, inter alia, by the Court action which the French Government has brought against the Commission for adopting the Terminal Equipment Directive under Art. 90 (3) EEC Treaty.

The third political factor shaping European telecommunications policy-making procedures and determining the outcome is the Commission's commitment to the "new approach" to technical standardization described above (sub. 3).

4.3 Political Consequences

On the political level, the legislative activities of the EEC in the telecommunications law field will eventually result in a shift of emphasis from national to EEC telecommunications law. European telecommunications law and national legislation implementing the EEC telecommunications law will be increasingly intertwined. The use of "framework directives" establishing general principles to be implemented, in more detail, by technical standards which are adopted by European standardization bodies, could result in a further deparliamentarization of telecommunications policy-making.

In spite of the obvious economies of scope resulting from the creation of a single European Market for telecommunications services and equipment and the "costs of non-Europe" which would result from a failure to implement the European telecommunications policy, it should not be overlooked that the creation of a Community-wide telecommunications market does not exclusively result in advantages for all participating political actors and market participants. Rather, the legal and political factors sketched out above would appear to result in considerable political costs which

- the Member States,
- the European citizens, telecommunications users and private operators alike,and
- third country enterprises

may have to pay as a price for the evolving European telecommunications market.

The supremacy of the Community legislation in the telecommunications field will eventually reduce the relevance of national telecommunications legislation. Although the gradual shifting of legislative powers with regard to telecommunications has only just begun and both its legitimate scope and its exercise are still controversial, a centralization of telecommunications regulation on the Community level is conceivable.

It could result in reducing the function of the national legislators to the mere "notarization" or rubber-stamping of Community legislation. In view of the merely consultative functions of the European Parliament, the democratic deficit of such a centralization of telecommunications legislation is obvious.

Secondly, the two-tiered regulatory process which results from the specific legal force of Community law directives will conceivably result in the development of a new administrative network between EEC and national telecommunications authorities.

These inter-administration networks will evolve to the extent that EEC directives in the telecommunications field will be implemented by the national bureaucracies, in particular the newly created regulatory bodies (p.e. OFTEL in the United Kingdom, the Federal Ministry of Posts and Telecommunications in the Federal Republic of Germany).

The organizational and procedural interfaces within these bureaucratic networks and the role of the Commission will require careful definition on the basis of Community law.

The obvious risks of the development of inter-administrative networks between the Commission and national bureaucracies are

- gradual erosions of legally defined competences,
- intransparency of decision-making procedures,
- use of "log-rolling" strategies between the Community and the national bureaucracies and
- erosion of the procedural safeguards of national administrative procedures (p. e. right to a hearing, right to a reasoned opinion).

Thirdly, the Member States' reluctance to delegate regulatory powers in the telecommunications field to the Commission and the "new approach" to technical standardization could result in an increased transfer of "political" decisions of telecommunications matters to "non-political" institutions, such as standardization bodies. By delegating powers upon European standardization bodies, both the Member States and the Commission appear to "win": The Member States can avoid transferring legislative and policy-making powers upon the Commission. The Commission, in turn, can achieve some form of European-wide decision-making and can assume a certain amount of control over it.

In spite of what might be preceived as a "deregulation" of telecommunications policy-making (namely: the transfer of decision-making powers upon European standardization bodies rather than decision-making by national government or Community organs), the political costs of the ensuing depolitization of decisions relating to telecommunications must be taken into account. If and when technical standards *replace* legislative decisions or if legislative decisions are too vague and imprecise to "program" the implementing technical standards, then standardization bodies are assigned a *political* role, which they are neither empowered nor equipped to play.

If these costs of the European re-regulation are to be avoided or , at least, minimized, the use of both the regulatory instruments and the powers to regulate telecommunications on the European level must take into account the regulatory experiences gathered in the Member States and build its regulatory structures upon the common principles of administrative (telecommunications) law.

References

Bebr, G. (1981), *Development of Judicial Control of the European Communities*, The Hague et. al: Nijhoff pp. 584-588.

Beyerlin, U. (1987), "Umsetzung von EG-Richtlinien durch Verwaltungsvorschriften?", EuR, pp. 126-145.

Boulois, B. (1979), *L'applicabilité directe des directives*, RMC, pp. 104.

Dashwood, A. (1978), "The Principle of Direct Effect in European Community Law", *16 Journal of Common Market Studies, pp. 229.*

Easson, A.J. (1979), "Can directives impose obligations on individuals?", European Law Review, pp. 67.

Ehlermann, C.D. (1987), "The Internal Market Following the Single European Act", CMLRev, pp. 361.

Ehlermann, C.D. (1981), "Die Verfolgung von Vertragsverletzungen der Mitgliedstaaten durch die Kommission", in Grewe, W.G. und Rupp, H. et al. (eds.), *Europäische Gerichtsbarkeit und nationale Verfassungsgerichtsbarkeit - Festschrift für Hans Kutscher*, Baden-Baden, Nomos.

Evans, A.C., (1979), "The Enforcement Procedure of Art. 169 EEC", European Law Review, pp. 442.

Ficker, H.-C. (1973), "Zur Angleichung der Gesellschaftsrechte in den erweiterten europäischen Gemeinschaften", in Fabricius, F. (ed.), *Law and International Trade - Festschrift für Clive M. Schmitthoff*, Frankfurt, Athenäum Verlag, pp. 157.

Fuß, E.-W. (1965), "Die Richtlinie des Europäischen Gemeinschaftsrechts", DVBl, pp. 378 seq..

Goldmann, B. (1969), "Le Rapprochement des Legislations", *Les Novelles - Droit des Communautés Européennes*, Brussels, pp. 883.

Ipsen, H.-P. (1965), "Richtlinien-Ergebnisse", in Hallstein, W. und Schlochauer, H.J. (eds.), *Zur Integration Europas - Festschrift Ophüls*, Karlsruhe, Müller, pp. 67.

Marburger, P. (1982), "Die gleitende Verweisung aus der Sicht der Wissenschaft", in *Verweisung auf technische Normen in Rechtsvorschriften*, Berlin, pp. 27.

Oldekop, D. (1968), *Die Richtlinien der EWG*, Göttingen, pp. 140.

Oldenburg, A. (1984), *Die unmittelbare Wirkung von EG-Richtlinien im innerstaatlichen Bereich*, München, Florentz

Ortlepp, B. (1987), *Das Vertragsverletzungsverfahren als Instrument zur Sicherung der Legalität im europäischen Gemeinschaftsrecht*, Baden-Baden, Nomos

Rambow, (1968), "Probleme bei der Durchführung von Richtlinien der EWG", DVBl, pp. 445.

Schenke, W.-R. (1980), "Die verfassungsrechtliche Problematik dynamischer Verweisungen", NJW, pp. 743 - 749.

Seidel, M. (1983), *Direktwirkung von Richtlinien*, Europa-Institut der Universität des Saarlandes, Saarbrücken.

Seidel, M. (1985), "Die Direkt- oder Drittwirkung von Richtlinien des Gemeinschaftsrechts", NJW, pp. 517 seq..

Zuleeg, M. (1980), "Die Rechtswirkung europäischer Richtlinien", ZGR, (474-477).

Zuleeg, M. (1969), "Das Recht der Europäischen Gemeinschaften im innerstaatlichen Bereich", KSE Vol. 9, pp. 261-294.

Comment Part IV

W.R.B. Wigglesworth

Oftel
London

In this session we have three contrasting papers, raising a range of issues from which it is difficult to extract a common theme; though the situations the papers address are certainly thought-provoking.

I take them in the order of the programme.

Changing Regulatory Policies: The Danish Case

Here is the description of a telecommunications country, as befits the home of Hans Anderson and Leggoland (the equivalent of Mecca to my young son) where all is well. One of the highest standards of living in the world is reflected in high telecommunications usage, with consequently favourable effects on the economics of telecommunications provision. The telecommunications system is claimed, and appears, to be extremely efficient and well structured.

The, by European standards, quite small regional operating companies described in the paper are exactly the sort of structure that we in the UK have tended to regard as the ideal, if it turns out that the scope for sustainable local network competition is limited to main business centres only. We have called this "a mosaic of ownership" (envisaging quite a number of moderate sized operating companies in a country the size of the UK); and we have envisaged such a structure potentially developing from the integrated cable television and telecommunications companies which are being established in the UK. The advantages we would see in such a structure, compared with the monolithic network monopoly of the previous 70 years in the UK, would be the potential for competition by emulation, or "yardstick competition", between local network operators and the opportunities created for diversity in innovation of the sort that we have seen in UK's only existing independent local network in Hull. It is therefore particularly interesting, from a UK perspective, to observe a system of separate operating companies in Denmark apparently operating extremely effectively.

Presumably, the rather relaxed approach to the development of competition in the customer premise equipment and services markets also reflects a history of past effectiveness in conditions of some diversity; in contrast to the all-embracing apparatus and services monopoly in the UK up to 1980, and the correspondingly strong reaction in the UK thereafter.

I found it difficult to follow the description of the regulatory arrangements; and I felt I could have done with a fuller treatment of this aspect. However, I suspect that it would be rather like trying to explain the unwritten British constitution. The subtleties are too difficult to explain except at inordinate length and usually in terms that are unintelligible to anyone but a native.

The success of the Danish system certainly does seem to reflect cultural differences and a Scandinavian genius for social organisation for which we in the UK have reason to be grateful in some aspects of our own cultural heritage.

I would certainly agree that the degree of efficiency of existing operations is a factor to be taken carefully into account in deciding on the extent and pace of the introduction of further competition, particularly network competition. At OFTEL we do not argue that network competition is desirable without limit. We see this as primarily a cost-benefit question. The loss of economies of scale and the further costs of duplication, arising from the introduction of a competing network, need to be balanced against the offsetting benefits in the creation of incentives to operate more efficiently, with lower costs for a given level of service, and in incentives to provide better services. It is consistent with this approach to consider the issues particularly carefully where existing performance is very good.

Nonetheless, my own concern would be with the long-term effectiveness of local monopolies, however well structured, in providing the best mix of service solutions for all types of user as the range of technical options for fixed links and mobile applications multiplies with technical advance. Will monopoly network providers keep up with technical and market developments to continue offering optimum services to all users? It is a tall order.

I would agree with the author of the paper that regulation aimed at promoting competition should, so far as possible, avoid imposing a market structure which prevents economies of scale and scope being exploited to the benefit of users. In the UK, with a few key exceptions where frequency availability has limited market entry, we have generally followed the philosophy of allowing network operators to compete in all sectors of the market, relying on accounting separation to prevent the suppression of competition in non-network markets. By the same token, I would be more sanguine than the author about the chances of effective regulation of this separation. The principle to be applied is that the network operator must treat his subsidiary ancillary business in exactly the same way as he treats its competitors. Though I do grant that the associated regulatory judgements can be quite difficult.

The issues arising from regulatory developments in Denmark are certainly of absorbing interest; and I have much sympathy with the open-ended final judgement.

Chances and Problems of Liberalisation of Telecommunications in Hungary

This paper is a complete contrast. Against the background of a country in economic crisis, facing the failure of successive centralised economic approaches, the telecommunications system is in a highly undeveloped state with very low penetration and an average waiting time for new service of 14 years.

Strong efforts are clearly being made, despite strictly limited financial resources, to develop the system. A whole range of new services (already familiar in the West) have been introduced during 1989; activities other than network operation are being liberalised; and options for restructuring the public network are under consideration as part of the reform programme.

The problems of developing the network are clearly formidable and the inclination from the paper appears to be to concentrate on extending the provision of telephone service on a monopoly basis. Given the scale of the problems one can certainly sympathise with this view; but I am not sure that it is conclusive. Next week a senior Hungarian delegation will be visiting OFTEL and I look forward to discussing the relevance of our regulatory experience to the Hungarian situation.

On the basis of this paper, my advice would be to see what might be turned to advantage in the Hungarian situation. The key point is that Hungary starts behind countries with more developed telecommuncations systems. But in today's circumstances this may be positively helpful. It means less sunk cost in out-of-date technology and less commitment to working practices associated with it. Development of a state of the art and cost effective network is likely to be much cheaper and quicker than it would have been previously.

How best to develop a network infrastructure that can be rolled out quickly in a way that might attract Western investment along similar lines to those being considered by Poland? Have a look, I suggest, at radio network possibilities, in mobile and fixed mode. Since radio is ready-made for competition, consider the possibility of allowing competing companies to develop the system. It is an approach that has worked well in the UK and could provide an example of market development for Hungary. If you conclude that at least the first tranche of such a development should be on a sole franchise basis, then keep open the opportunity of allowing a competing system in due course. This will be a powerful stimulus on to the first network provider. Look for a range of equipment and services suppliers for such a system.

Aim to establish a regulatory function that is open in its decision making.

Hungary ought to be a country that is likely to attract external investment, provided it can develop an appropriate market structure. It is an exciting prospect.

Regulatory Instruments and Powers to Regulate Telecommunications Services in Europe

This paper addresses a series of different regulatory issues concerned with not just one country but with the European Commission's approach to the development of telecommunications throughout the European Community.

As the telecommunications regulatory body within the UK, OFTEL's direct competence does not extend to international telecommunications policy issues, which are the responsibility of the Department of Trade and Industry. You will understand that I am, therefore, not in a position to comment officially on the UK position in discussions at Brussels; and points in this paper should be read as no more than the personal comments of an interested observer.

In fact I am not competent to provide a critique of Dr. Scherer's expert and authoritative analysis of the Treaty provisions under which the Commission is seeking to move Europe down the path of telecommunications liberalisation. The British Government has evidently found some difficulty in reconciling its support for the Commission's liberalisation objective with reservations about the means adopted for its achievement. To an extent, it seems to me essentially a political choice for the Commission to make of the means it considers most likely to lead to the achievement of that objective within a given timescale.

In the longer term, the most interest perhaps attaches to the regulatory implications of the European developments. The formulation of rules governing the progress of liberalisation clearly does have regulatory effects, though I doubt whether the significance of these can be fully measured without considering the extent to which the final regime will be subject to the rules of competition, the effectiveness of which is often underplayed in public comment within the UK.

I was particularly interested in Dr. Scherer's comment that the current process in Europe is leading to the "deparliamentisation" of telecommunications issues. Similar criticism has been made about the impact of the current regulatory arrangements in the UK. I would myself draw a distinction between the public discussion of regulatory issues and discussion confined to Parliament. Within the UK, at least, the setting up of specific telecommunications regulatory arrangements has led to a much more structured and open public debate of such issues than was the case when telecommunications was run by a state corporation. There is no bar to parliamentary debate on such issues, though the more effectively they are handled, the less there is likely to be a need for their debate in Parliament. I should have thought a similar distinction could be drawn at European level; though I accept that legal and administrative complexities may inhibit comment by all but the most well informed.

The paper does not examine the question of whether some more highly structured regulatory approach should be adopted at European level. My own view would be that, in current circumstances, where opinions remain divided on many policy options, it is better to follow a route of informal contact and discussion with those concerned in different Member States and beyond - of which this Conference is one example.

Part V

Convergence of Broadcasting and Telecommunications

11

Diverging Regulatory Scenarios for Integrating Broadband Communications

Emilio Lera

Telecommunications Division
Price Waterhouse
Madrid

1. Introduction

The ongoing advances in technology are proclaiming a qualitative change in the distant communication process by allowing full motion image interchange between communication subjects. These types of advanced communications will require availability of powerful telecommunication infrastructures providing broadband services to accessing, transporting, storing and interchanging any information which requires very large transmission rates. The type of information handled varies from large data files to high resolution graphics, suitable mainly for business sector operations, and image information suitable for some economic activities, mainly in the service sector, like education and tourism.

Person to person visual communication will be possible through these infrastructures, allowing an improvement in the communication process by complementing the voice with kinetic and prosemic information. Unfortunately, this improvement is not worthy enough for the users to compensate the cost increase upon the voice communication tariffs, which likely will be imposed by the exploitation of the new broadband infrastructures.

There is, however, a type of services which have a general and growing acceptance, which is the video entertainment diffusion services, mainly television. They are nowadays being supplied upon other infrastructures like cable distribution networks, direct broadcasting or shared telecommunication satellites and terrestrial emitters.

Technology allows provision of such services on the broadband networks, with some economic advantages derived from the avoidance of duplication of infrastructure investments, and with the extra advantage of an almost unlimited bandwidth available to locate video channels, in contrast with the trend towards spectrum saturation being experienced by the herzian diffusion media. In this way, there seem to exist techno-economic reasons to integrate both types of services, the telecommunications as well as the diffusion ones, on the future telecommunication broadband infrastructures.

However, there are important differences in the environmental conditions acting upon the dynamics of both services. To be mentioned are service provision conditions, market supply diversity, demand requirements, driving forces of technology utilization, and the applied regulation status. Even if these factors will evolve up to the moment when the new broadband telecommunication systems and equipments are available on the factories and the corresponding infrastructure can be launched, probably in ten to fifteen years time, the different dynamics of both types of services do not assure a convergence between these diverging factors.

In this paper we will describe the *differences in the regulatory factors* which could constrain the ways in which integration could take place. Some forecasts of possible scenarios of the future applied regulation will be presented, together with the likely impacts on the integrating policies. These factors are the result of a survey upon the European countries; they are not representative for any specific country. Rather, each country has its exceptions, which do not change, however, the overall regulatory framework outlined in the following.

2. Systems Organization

It should be interesting to start the discussion by reviewing the *characteristic elements which define the respective systems*, i.e. those of the telecommunications system and those of the broadcasting system, especially in the television segment. It is clear that these systems justify the different regulations being dictated to create, protect and develop them.

The particularities of both types of services have defined the actual status of the respective systems. The universal service concept, introduced by T. N. Vail, first President of AT&T, in 1907, has justified the definition of a single policy, the implementation of a powerful organization, and the objective of service uniqueness[1] in the telecommunications sector. These elements are the basis of the corresponding National Telecommunication System, i.e.: a policy, an organization, and a service. The provision of such a service required a powerful and costly infrastructure, which had important economies of scale to an extent that discouraged potential competitors.

The weight of such an infrastructure together with the universal character of the service, originated in most of the countries a monolithic and monopolistic organization which defined by itself the policy and guaranteed the service uniqueness. In Europe this organization had the character of a public entity which together with the fact that being an important net income generator, assured the stability of its legal status.

The availability of that infrastructure, technologically complex, has *linked the telecommunications service policy with the telecommunications industrial policy*, in such a way that the first changes in the basic regulation were aimed to stimulate the industrial markets, as one can see e.g. by the liberalization of terminal equipment.

Television services, because of their mass communication character and their large social penetration, had the potential capability of acting upon social cohesiveness. Their message content could influence the diffusion of ideologies and the moral and cultural integrity of the population. In this way, the European countries had defined such services to be of public interest, creating a monopolistic organization to broadcast them. These circumstances have *linked the television broadcasting policy with the information and communication policies*, imposing a severe control upon the access to the media. The first changes in the regulation experienced by some European countries, allowed the creation of private broadcasters but with a rigorous control of the promoters and their programm sources, as can be seen e.g. by the creation of IBA and ITV in the United Kingdom.

---

1 By "uniqueness" we mean a concept that there is not any difference in service provision conditions all over the country (i.e.: user to network interface, service facilities, tariffs, quality of service, etc.). Contrary to this we believe the concept of "universality" to apply mostly to the everybody right of access and usage of service but not necessarily under the same conditions.

In the United States, the First Amendment assured the total liberalization of the audiovisual services, which allowed a multiplication of the national and local television chains, without any kind of constraint neither in contents nor financial sources, leaving it to the market to regulate the sector.

In spite of these different links existing in both types of services, they have followed similar developping patterns in Europe, thus having achieved recently a *common status, which is: State reserved services provided through public entities under a monopolistic regime*. There must be remarked, however, *peculiarities affecting these developping patterns*, mainly in Europe:

- The *service regulation basis* is the uniqueness of supply for the telecommunications services and the conservative message content for broadcasting services.

- The *utilization of the technical facilities to provide other agent's services* is restricted, by imposing special conditions in both types of services.

- The *implementation of new technical means by other agents to provide services* is forbidden in telecommunications, while it is restricted in television broadcasting to local emitters.

- The *financing mechanisms* used for telecommunications services are based on the billing of usage of infrastructure by means of State regulated tariffs. Those used for broadcasted services are based on the terminal ownership and publicity.

- The *driving forces* behind the developing patterns are the technology application and the diffusion of entertainment, respectively, both related with the above mentioned links.

- The *market orientation* of the telecommunications services is the supply, while that of the television services is a standard demand.

In both cases, regulation is applied to maintain the status of the system by protecting the providing entities from any intrusion into their business.

Technology and market dynamics, however, have disturbed the stability of the systems by allowing provision of services through technical means peripheral to those of the monopolistic infrastructure and by outsiders of the licensed entities. There are the cases of television distribution by satellite and the enhanced communication facilities implemented in terminals and PABX, which allow other agents and users themselves complementing the public offerings.

These changing situation has forced important regulatory changes on both sectors but with non converging aims.

3. Service Requirements

Before analysing these different regulatory elements it should be worthy to review those differences in the service requirements which may influence the particular regulatory policies.

The following *operating characteristics of both systems* are well known:

- Provision of a unique bandwidth (4 KHZ) in the actual telecommunications (TC) infrastructure,
- Digital transmission rate limited upon a bandwidth (64 Kbps), not enough to provide video services,
- TC tariffs are based on distance and usage time with a trend to be cost related, while TV herzian distribution use to be free and cable network distribution has a flat rate.
- Mainly not more than one simultaneous TC service provision per network terminating point (residential user), while two to four simultaneous video channels are consumed.
- Very short service provision duration (very few minutes) against hours of continuing TV consumption.
- On demand TC service provision in contrast with many hours of continuous, consumer independent TV service provision.
- Hierarchical and meshed TC network structure to optimize the full network connectivity cost, compared with compound star TV terrestrial herzian network and tree cable distribution network structures.
- Very complex TC network control procedures (signalling, control and operation), but very simple ones in TV networks.

- Limited TC service availability due to routes' (and some systems') congestion, unexisting problem in TV distribution.

- The very costly TC infrastructure, paid by the generated income of a single service (telephony) is necessary to provide other advanced TC services, a requirement which until today does not exist in TV distribution networks. In other words, economic viability of the corresponding networks has not the same socio-economic importance.

These distinct service requirements have their weight in the corresponding regulations being adopted.

4. Regulation Policies

Once we have reviewed the system organization of both types of services and the differences in service requirements which may influence the corresponding policies, we are prepared to analyse the existing regulatory divergences which may constrain any service integrating policy upon the future broadband networks.

We may start by considering the *legal character of both types of services.* Some TC services hold the condition of public services, which implies that special provision requirements are imposed to the providers, like service extension objectives, approval of tariffs, certification of terminal equipment, service accessing clauses, etc. However, these conditions are not applicable to new services open to competition, which originates a mix of legal status in the sector, which is of relevance to the providing entity of the public services. In some countries TV broadcasting services have also this public service character, with a clear trend to remove it in the new regulation being dictated; it must be noted that a mixture of provision conditions does not exist.

The new TC system organization being created in Europe establishes as a basic principle the *separation of the regulatory functions from the network operating functions,* which used to be in the hands of the same entity. In TV distribution services there is not any special requirement about the location of the regulatory function. Very often a distinction is made in the case of cable distribution licenses between the license to lay out cables and the TV distribution concession; it should be possible in these cases that the PTT plays the double role of granting the distribution network layout license while simultaneously participating in its installation and exploitation.

It is to be noted the existing restriction in the USA to "common carriers" to participate in the audiovisual diffusion business in their operating area, even in collaboration with other agents. However, this regulation will evolve as they are already allowed to provide services in unattended areas, and they may rent infrastructure distribution capacity to independent providers. Technology could force this regulatory change, if distribution is considered as a part of the integrated broadband service business; some actual experiments are representative for this new situation. The RBOCs penetration in the European cable distribution segment (mainly in the United Kingdom and France) shows the interest of these companies to prepare themselves for a new regulatory framework by taking over the business know-how where it is available.

The *definition and control of service provision conditions* are State responsibilities in the case of TC services; concessions for the provision of services are then granted by a national authority. However, TV broadcasting concessions tend to be granted by regional or local authorities, with minor interference from the national entity, the latter taking place only in the assignation of frequency channels for herzian TV channels transportation and diffusion. In the above-mentioned situation of requiring a dual concession for laying out and exploiting cable distribution networks, the layout license is granted by the national authority while operation and commercial concession is granted by a local authority. It must be noted that up to now there is never such dual responsibility in telecommunications.

A difference with respect to the USA situation must be noted. The cable networks are classified by regulation as interstate, preserving them from the States' authority in granting licenses. A similar situation is present in the Federal Republic of Germany concerning the rights of the Länder.

The public character of TC services has forced the legislator to establish *limits to the participation of foreign capital* in their infrastructures. Even if nowadays there is a trend to allow private capital participation in the licensed entities, this participation is being constrained in such a way that never a single company could achieve the control of the entity. These limits are not, in general, applicable to entities providing non basic services, a fact which may favour the future concentration of this economic segment. In TV distribution the situation varies depending on the technical media used. In general, there are not such limits to foreign participation in cable networks, but they exist in the herzian transportation network; new licensed TV channels, even if under private ownership, have also this constraint.

The new TC system regulation is being dictated to promote competition in the provision of non basic services. This objective obliges to assure full ***access to the networks for third agents*** in order to implement and provide their services; the EEC is defining the European Telecommunication Norms (NETs) to care about this requirement. Cable distribution networks are not subject to such a regulation, a fact which may allow a monopolistic provision of enhanced services upon these infrastructures. It is not clear yet if the general TC regulation should be applied to grant concessions for these kind of provision, as the cable exploitation licenses are to be granted by a different authority and potential offering of these services is constrained to a limited geographical area. In those cable networks which are allowed to have TV production capabilities, sometimes a requirement is imposed to permit part-time access to a TV channel for the minority groups, but only to diffuse their messages, i. e. without business orientations.

The changes being introduced upon the TC system do not modify the general objective of ***network integrity***. Even if any service should be open to competition, requiring in some cases the implementation of value added networks, these are obliged to comply with defined standards so as to ensure that integrity. Distribution networks are not subject to this requirement, which certainly will not favour their interconnectivity and compatibility so as to establish a conglomerate of interconnected local distribution networks as an alternative to other infrastructures. It may be that this situation was provoked to protect the stability of the TC system.

Another difference with respect to the US situation may be remarked. Since the Radio Act of 1927, electronic media are not subject of First Amendment protection because of the limited spectrum resource available. A consequence is a constraint to operate DBS satellites, thus requiring two licenses, one for building them and another for launching and operating them. This fact could originate a distortion to the TV satellite distribution business.

Standardization is playing a mayor role in the new TC system, as being the instrument to assure network integrity and to avoid network fragmentations (others than those based on the operating ambit like local, interurban and international). Uncontrolled fragmentation could beat the system stability breaking up the uniqueness service objective of the system. *Type approval and certification* of network systems and peripheral devices take then an increased importance, so as to regulate a functional separation of these activities from those of network exploitation and commercial activities. Up to now there is not a similar relevance of the standards in the implementation and planning of cable distribution networks; neither kind of approval is exerted for network equipments or peripheral devices (except conformance with the user security requirements).

To assure economic viability of the providers of public services, some (basic) *services are reserved* for their exclusive provision, like telephony or telex services. This regulation may generate conflicts in an integrated broadband environment where other services are provided in competition, or whenever the users could subscribe the services in bundles. This reservation does not exist in diffused TV services, only some contents are reserved to public channels like political electoral publicity.

TC public services are subject to *tariff regulation*, but they are not the non-basic services. In certain countries cable TV distribution needs regulatory tariff approval, but there never exist two distinct regulatory regimes; it could happen that this difference will appear as soon as some value added services will be provided through such infrastructures, like teleshopping or telesecurity. To allow for a fair competition in the provision of non basic services, regulators are forcing to adopt a *cost related TC basic services tariff pattern.* Up to now, there is not this mixed services provision in cable networks; however, flat tariffs are used, which are more related to the service acceptance by the market than to the infrastructure operating costs, a situation which could originate conflicts if infrastructure would be rented to other providers of value added services.

Some special regulations must be mentioned being applied to distribution services, especially to cable, which are not applicable to TC services. TC licenses, even for non basic services, are not constrained to provide service to a *limited geographical area.* Moreover, they are not obliged to *provide access to specific information sources* (obliged distribution of public TV channels), being a matter of these sources to comply with defined standards and to request a connection to the infrastructures to allow their access. Neither, up to now in many European countries, they are allowed to *resell spare transmission capacity* to third agents (distribution of pay TV channels); this reselling is allowed to basic service providers, for instance with the implementation of a basket value added services tariff. Neither they are obliged to *provide free service to minority groups* (obliged access permission to minority groups to diffuse their mesage); however, in some cases there are special tariffs for some social groups like elder people. Nor they are requested to *pay copy-right fees* for the transported information (program and channel copy-right fees paid by cable distributors even if they do not produce or assemble the channel programms).

5. Evolving Regulatory Scenarios

The above regulatory situation is changing in most of the countries to adapt it to the new environment being created by the technology and market dynamics. We may draw some likely regulation scenarios which could favour or constrain the potential integration of both types of services.

First Scenario: ***Planning and Operation of broadband infrastructures will likely be reserved to a large service provider***, regardless of the services which have been reserved to it at that time. To be consistent with the existing regulatory principle of assuring the economic viability of the network operator, ***a minority of broadband basic services will likely be reserved to the network operator***, probably only videotelephony which has the voice signal in common with the actual reserved services. It could happen, however, that broadband services be provided in competition from the moment they are introduced, which would leave it to the market to drive the services penetration. The ***service uniqueness*** would be in question as service providers should try to differentiate their offer by innovating in the provided services. Integration of TC and TV distribution services should be, in this case, a matter of service provider's market approach, with many different situations in the market place.

Second Scenario: It could happen that the current liberal ideologies, mainly in Europe, considering the character of that infrastructure of being a new implantation, promote a regulation which would ***open up to competition the network laying out and its exploitation.*** This new business opportunity will certainly produce a network fragmentation, with very strong difficulties to maintain the actual objective of ***network integrity***, as infrastructure operators could adopt the most advanced technology to improve their network performance, without taking care very much of existing ***standards.*** This should produce also an industrial market fragmentation, as usually standards were also oriented to maintain the integration of these markets. The integration of TC and diffusion services should be in this case a matter of market receptivity, being not the common driving force for all broadband networks.

A Third Scenario could be developed: It could be possible that in many countries existing operators of cable distribution networks, mainly those with a network structure which supports interactivity, will be allowed to provide value added services. They will likely start by providing image based services, like teleshopping or on-demand television, but if economy justifies it they will continue by providing services with a large potential demand, likely in competition to other providers who utilize the TC network. This situation should produce a confusing market structure, which certainly could not favour the integration of TC and TV distribution services upon the future broadband networks.

A common effect to these scenarios comes up from market growing. The traditional self-financing of the TC network operators to enable the network modernization and introduction of new services, could bring about problems to develop a market for services the demand of which is strongly dependent on their existing penetration at that moment, i.e. those in which the user acceptance is related to the capability of communications with other users, like it should be the case with the videotelephony. The limited resources to implement a large extended network in a short time, in such a way that demand satisfaction could not be constrained by network coverage, would surely phase the network development according to demand materialization and its new generated income. This situation would smooth the service penetration rate. Unless *policy makers would provide financial mechanisms* to modify the above situation the economic viability of broadband networks should not be achieved in a short time, which would not favour integration of services . From allowing participation of foreign capital up to total deregulation, passing through State financing, or tariffs deregulation, many policies are possible. We will likely be witnesses of the transformation of the existing TC system organization.

6. Conclusions

If we accept the principle that from the techno-economic point of view it is desirable to integrate all types of communications (in this case point-to-point and point-to-multipoint services) in a single network, to analyse the viability of this integration we must consider the existence of important peculiarities in the system organization of both types of services. This differentiation of systems, together with distinct service requirements, have originated particular regulations in the two sectors, which would influence the way in which that integration could take place.

The existence of different technological options to be adopted by the network operators and service providers, as well as the different market dynamics acting upon each service type, would produce a divergence in the regulation evolving patterns. Several modifications in the respective regulations are possible, each one with its own consequences for the sector's evolution and integration trends.

To avoid the appearance of irreversible situations, at least in the long run, it will be necessary to *devise and implement policies aimed to assure the desired integration.* However, the European policy makers are acting independently upon each sector, which could not assure the convergence in the above-mentioned evolving patterns as well as the creation of a community of interest about that integration between all the actors. In fact, the implementation of some coordinating policies could create a set of conflicts between these actors, as each one need to give up part of its actual business position to participate in the new business opportunities. Regulators, infrastructure operators, TC service providers, manufacturers of terminal equipment, audiovisual distributors, and even regional politicians, are not minded to accept these conflicts.

It must be emphasized that integration is not mandatory. From a pure economic point of view it could be innefficient, as the market's fragmentation produced by a global deregulation of both sectors should create a market growing, which could benefit the industry, the labour force, the technical and social innovation etc. We may regard what is happening with the informatics sector, its dynamics, systems renovation, and economic growth, to question the suitability of the TC services integration. However, we think that if this technology push is as successful as it used to be, the integration capability of IBC technologies will be used, and integration shall be achieved after a certain transit time. Stability of the actual TC system should favour this trend.

12

The Convergence of Broadcasting and Telecommunications in Japan

Tatsuro Hanada

Research Institute of Telecom-Policies and Economics (RITE), Tokyo

1. Introduction

The convergence of broadcasting and telecommunications is a very important and also interesting theme when we consider the impact of information and communication technologies in the fields of politics, economics, society and culture. Throughout developments in this convergence, we find complex interdependency of various factors.

In discussing this subject, we need to take the following points into consideration:

- Convergence is to be considered not as a duet of broadcasting and telecommunications, but as a trio of broadcasting, telecommunications and information processing, i.e. computer technology.

- The process of convergence, which starts in the field of technology, on the one hand leads to convergence and on the other hand to divergence of economic and then institutional aspects.

- Convergence takes a different form in different countries depending upon the legal and political systems and the historical experiences of each individual country.

The technological convergence of computer technology and telecommunications was one of the determinant conditions for the Japanese telecommunications reform in 1985. There were also pressures from both domestic and international actors with political and economic interests. The convergence of broadcasting and telecommunications is not simply a phenomenon brought about as a result of technological development but also one of the consequences of structural changes in the telecommunications regime. Through the liberalization of both facility-based and non-facility-based markets in telecommunications, the conventional order of broadcasting transmission networks is undergoing fundamental changes. The convergence of broadcasting and telecommunications in Japan seams to be focused in the long-run on a broadcast reform in a liberalized environment. It is a process driven by techno-economic factors in which an old framework is gradually replaced by a new one accompanied by marketplace restructuring and institutional conflicts.

This paper presents an analysis of the following aspects in Japan:

- the legal and regulatory framework of broadcasting and telecommunications in which the underlying structure of current conflicts has been forged;
- factors driving the convergence;
- major issues in policy making with regard to convergence and the recent political process;
- theoretical considerations relating to the functional definitions of broadcasting and telecommunications and the interpretations of the process of convergence which follow from them;
- industrial and regulatory consequences of the convergence of broadcasting and telecommunications, and the responsibility for consolidating cultural and social values confronted with techno-economic developments.

2. Background of the Convergence of Broadcasting and Telecommunications

2.1 The Existing Framework of Broadcasting and Telecommunications

Before discussing the subject of convergence, we need to look at the legal concepts of broadcasting and telecommunications and their current regulatory frameworks.

2.1.1 Legal Definitions

2.1.1.1 Telecommunications

In the Telecommunications Business Law, "telecommunications" is defined as "transmitting, conveying or receiving codes, sounds or images by wire, radio or any other electromagnetic method" (Article 2, paragraph i), and "telecommunications service" is defined as "intermediating communications of others through the use of telecommunications facilities, or any other acts of providing telecommunications facilities for the use of communications of others" (Article 2, paragraph iii). "Telecommunications business" means "business which provides telecommunications service in order to meet the demand of others." Notably, businesses related to wire radio broadcasting, wire broadcast telephon service, cable television broadcasting service, and the acceptance of applications for the use of the cable television broadcasting facilities are all excluded from telecommunications business under the provision of article 2, paragraph iv.

"Intermediating communications of others" is interpreted as completing communications of others remotely located in response to their request, without any change in content, regardless of whether it is one-way or two-way, or is performed by means of machine or human labor.[1]

Furthermore, the law stipulates prohibition of censorship (Article 3) and protection of secrecy (Article 4) concerning communications handled by a telecommunications carrier, in compliance with Article 21, paragraph 2 of the Constitution of Japan. In other words, the secrecy of messages conveyed by telecommunications is upheld as a fundamental principle.

To summarize the above, telecommunications consists of the exchange or distribution of information between particular persons and the content of this information is left as it is

---

1 (In Japanese) Commentary for the Telecommunications Business Law, 1987, p. 14.

independent of the telecommunications business.

2.1.1.2 Broadcasting

"Broadcasting" is regulated by the Radio Law and the Broadcast Law, and is defined as "transmission of radio-communications intended to be received directly by the general public" (Article 5, paragraph 4 of the Radio Law and Article 2, paragraph 1 of the Broadcast Law). "Broadcaster" is defined as a "person engaged in broadcast enterprise who has been granted a license for his broadcasting station in accordance with the provisions of the Radio Law" (Article 2, paragraph 32 of the Broadcast Law). A broadcaster, who is responsible for the contents to be broadcast, i.e. a broadcast program, must initially obtain a license to run a radio station for the purpose of broadcasting.

Broadcasting, according to this official definition, focuses on transmission, rather than reception of information using radio waves. While broadcasting aims at direct reception by the general public, it is not required that all of the unspecified number of viewers and listeners receive the information and no particular receiver of the information is assumed in the law.[2] Thus, broadcasting is "a concept which focuses upon the unilateral supply of information" and is not considered a form of telecommunications.

"Cable television", which is intentionally excluded from the telecommunications business in the Telecommunications Business Law, is defined as "transmission of wire telecommunications aimed at direct reception by the general public" (Article 2, paragraph 1 of the Cable Television Law) by analogy with the definition of broadcasting. While broadcasting in a narrow sense (i.e., broadcasting by radio wave) is regulated by the Radio Law and the Broadcast Law, cable television is regulated only by the Cable Television Law. However, in this law "a cable television operator" and "a cable television broadcaster" are conceptually distinguished from each other so that channels can be leased.

From what has been described so far, broadcasting in a broad sense (whether it is by radio wave or by wire) may be summarized as the transmission of telecommunications for unilateral distribution of information to an unspecified number of people and also as a sphere where the business is concerned in the content of the information.

The provision for subscription broadcasting service (payTV) was introduced with the revision of the Broadcast Law and the Radio Law in 1988, and the provision for broadcasting service by way of "telecommunications" satellite (hereafter referred to as

2 Ibid., p. 15

"CS") was introduced as a result of the revision in 1989. These matters will be discussed later in the section regarding the main issues in policy making.

2.1.2 Administrative Regulation

2.1.2.1 Telecommunications

The telecommunications business consists of three sectors which are subject to different levels of regulation. First of all, the industry is divided into Type I telecommunications, which provide telecommunications service by establishing telecommunications facilities", and Type II telecommunications, which are defined as "any telecommunications business other than Type I". In general, Type II telecommunications provide service based on telecommunications circuit facilities leased from Type I telecommunications carriers. These are further divided into "General Type II" and "Special Type II" telecommunications business depending upon whether the business is provided to unspecified or specified users.

What is characteristic of the above classifications is that Type I business provide both telecommunications facilities and services, whereas Type II business provide only telecommunications services, which are separate from facilities management.

As for regulation of these businesses, a Type I telecommunications carrier is required to obtain permission from the Ministry of Posts and Telecommunications (MPT) before it commences, suspends or discontinues its telecommunications business and also must obtain authorization for a tariff which sets forth provisioning conditions of services including charges. Moreover, a Type I telecommunications carrier must not, without due reason, refuse to provide telecommunications service. MPT does not grant permission to any person who does not have Japanese nationality; any foreign juridical person or association; or any juridical person or association with more than one third of total voting rights held by such persons or bodies.

Before commencing Type II telecommunications business, a Special Type II provider must obtain registration from MPT and a General Type II provider must notify the Ministry. Both types must also submit notification before suspension or discontinuation of their business. Notification must also be submitted concerning tariffs of Special Type II providers. There is no obligation to provide telecommunications service, and no (juridical) person or association is disqualified because of foreign nationality.

2.1.2.2 Broadcasting

Here we look at major regulations concerning conventional terrestrial broadcasting and cable television.

As to terrestrial broadcasting, any person who intends to commence broadcasting business is required to obtain a license for a broadcasting station as provided for by the Radio Law. The licensee must comply with the following conditions: he must have a close relationship with the region he serves (human and capital linkage to the local community); he must not attempt to exercise concentrated control of mass media (no single individual may concurrently own or control radio, television and newspapers); and he must not have a foreign status (i.e., a person or juridical person with foreign nationality, or a juridical person with more than one fifth of the total voting rights held by persons or bodies of foreign nationality).

Furthermore, the Broadcast Law regulates broadcast programs. The regulations

- include codes for compiling of broadcast programs (for example, these (i) must not disturb public security and good morals and manners, (ii) must be politically impartial, (iii) must broadcast news without distorting facts, and (iv) regarding controversial issues, must clarify the issue from as many angles as possible)

- stipulate the principle of harmony among the broadcast programs (application of this principle is restricted to television broadcasting in general and radio broadcasting by NHK (The Japan Broadcasting Corporation))

- and oblige each broadcaster to establish a standard for compiling of broadcast programs as well as a consultative organization on broadcast programs.

Concerning Cable Television, in compliance with the Cable Television Broadcasting Law, any person who wants to start cable television broadcasting on a scale of more than five 2hundred service terminals must obtain permission for construction of facilities from MPT, while only notification is required for construction of cable television facilities with fewer service terminals. As in the case of broadcasting, foreign nationals are excluded from licensing.

A cable television operator is a person who conducts business through the usage of cable television facilities by submitting notification for his business as well as for a tariff to the MPT and who is also under the obligation to provide services. A cable television operator is obligated to retransmit terrestrial programs addressed to its broadcasting area but must not retransmit distant programs without the consent of the broadcaster. When the operator is compiling programs by himself, he is subject to the same regulations as in the case of terrestrial broadcasting.

2.1.3 The Characteristics of the Japanese System

The characteristics of the Japanese system for broadcasting and telecommunications can be summarized as follows:

(i) Responsibility for the network and for the information content (programs) has been unified in the Japanese broadcasting system. Under the "unification model" of the two elements, hardware and software, broadcasting enterprises have to construct and operate their own transmission facilities (broadcasting networks) under licenses issued by MPT in order to provide programs to their audiences. While the transmission network facilities are regulated by the Radio Law, broadcast programs are regulated by the Broadcast Law. This principle is applied not only to wireless broadcasting but also to cable television. In other words, cable television is included in broadcasting in a broad sense and thus it is called "cable television broadcasting".

This is a big difference compared with, for example, the case of West Germany where the "separation model" of technical and cultural elements is applied under the federal system of the state. As is examined later in this paper, the unification model for broadcasting in Japan is now being challenged due to the convergence of broadcasting and telecommunications.

On the other hand, in the field of telecommunications, a new system was introduced by liberalization in 1985 to distinguish a network operator who constructs and operates a transmission network from a service provider who conducts his business from leased facilities.

(ii) The concept of broadcasting, according to the legal definition, is composed of the following four elements : 1) reception by "the general public"; 2) "direct" reception by the general public; 3) the "intention" on the side of a broadcaster of direct reception by the general public; and 4) "transmission of radiocommunications" by a broadcaster. The concept of cable television is also composed of the same elements except that "transmission of radiocommunications" is replaced by "transmission of wire telecommunications".

(iii) No administrative regulation is applied to the contents to be transmitted by telecommunications, while in broadcasting there are various regulatory provisions concerning the contents to be transmitted. The services that are defined as broadcasting services are, therefore, subject to the legislative regulation of the Broadcast Law and to administrative regulation by MPT. If such services are defined as telecommunications, they are released from these regulations.

2.2 Factors Promoting Convergence of Broadcasting and Telecommunications

In the following, factors generating and promoting the phenomenon of convergence between broadcasting and telecommunications will be studied in more detail.

2.2.1 Structural Changes of Transmission Networks due to Technological Innovation

Conventionally, broadcasting service was provided through transmission networks for broadcasting, such as terrestrial waves for radiocommunications and coaxial cables for cable television, while telecommunications service was provided through telecommunications circuit facilities; within the telecommunications service, each type of service was provided through its own transmission network, in other words, each type of service was separated from others coupled to the corresponding transmission network.

However, as a result of the development of basic technologies such as semiconductor devices, digital technology, and fiber optic communications, transmission networks as a whole have undergone transformation on various aspects. This has contributed greatly to the technological convergence of computers, telecommunications and broadcasting. The structural changes of transmission networks can be classified in the following five categories:

- external and internal linkages between transmission networks and computers,
- digitization of transmission networks,
- transmission networks with larger capacities, multiplex technology and wider bandwidths,
- shared use of multi-functional transmission networks,
- integration and combination of different kinds of transmission networks.

This process started with structural changes in terms of quality and quantity and then gradually led to functional changes within transmission networks. Consequently the conventional framework of one transmission network for one type of service began to deteriorate and entered the process of fundamental transformation.

2.2.2 Diversification and Sophistication of Communications Needs

Supported by technological progress, diversification and sophistication of communications needs will advance the level of convergence. Diversification of communications needs leads to segmentation of the communications service market. The market is segmented into a number of submarkets. This is economically viable and network costs for a submarket decrease so dramatically due to technological improvements as not to require economies of scale. Both of these conditions have already been realized and a variety of services have been provided to a small group of people. For example, CUG (closed users group) services have been introduced into submarkets to satisfy the needs for distribution of messages to the members of a group.

In short, sophistication of communications needs requires sophisticated functions, namely, those of computers. The network has met needs for such communications in two respects. First, the need for value added services has been met by integrating computers within the telecommunications network. Second, videotex service was introduced into the market by connecting and integrating computers with the public switched network, thus achieving economies of scale and lower costs for individual users. The cost of information processing functions distributed among a large number of users through the public switched network by achieving shared use of computers.

2.2.3 Liberalized Environment of Telecommunications

The principle of competition was introduced into the telecommunications market by the Telecommunications Business Law which took effect in April 1985; at the same time, Nippon Telegraph and Telephone Corporation was privatized by the NTT Law. This reform was aimed at altering the state of ownership of transmission networks as well as the status of service providers in response to the above-mentioned structural changes relating to transmission networks and the diversification and sophistication of communications needs, which are giving rise to a transformation of the industrial order.

This liberalized environment has laid the foundation for convergence of telecommunications and broadcasting at a corporate level. More specifically, one cable television operator categorized in a broad sense as a broadcaster was licensed as a Type I carrier and started to provide telecommunications services such as leased circuit service. That first and so far only example is Lake City Cable Television (LCV) in Suwa City. Another impact of liberalization is the emergence of private CS operators. Among the various types of network operators that are licensed as Type I carriers are the two CS operators, JC-SAT and SCC. Earlier, it has been agreed to purchase CS from the U.S. according to the MOSS (Market-Oriented, Sector-Selective) formula, which was worked out in a bilateral negotiation with the U.S. at a high official level. Accordingly, the two newly licensed operators bought large size CSs made in the U.S. and successfully launched them by Arianne Rocket in March and June of this year. Applications of the transponders in both satellites have resulted in some changes in the Japanese broadcasting system.

2.2.4 Impact of "Telecommunication" and Broadcasting Satellites

CS has a sufficiently large capacity to supply a number of point-to-multi-point transmission paths. As a result of this distribution function, it is natural that the possibility of applications similar to broadcasting service is being considered. CS is also useful as distribution networks for cable television programs. Consequently, CS industry has come to overlap broadcasting industry in many respects. In order for CS to be classified as a transmission network for broadcasting service, the separation model must be introduced into broadcasting.

Based on the unification model, DBS has been developed and constructed by broadcasters with the financial support from the government. BS-2, which is now in operation, is financed by the government (40%) and NHK (60%). BS-3, which is scheduled to be launched in 1990, is financed by the government (32%), NHK and private broadcasters (59%), and by Telecommunications Satellite Corporation of Japan, a public corporation supported by MPT (9%). However, taking into account its financial burden and the high risk on the side of broadcasters, an inquiry commission set up in MPT proposed the establishment of an independent corporation for the procurement of BS-4. In this way the development of DBS has entailed introduction of the separation model.

Also on the technological front, it has been decided to develop BS-4, which is considered to be the most appropriate for high definition TV broadcasts, as a satellite for both broadcasting and telecommunications services. Thus, the distinction between broadcasting and telecommunications is disappearing in satellites.

2.2.5 Future Vision of ISDB

MPT presented in 1984 its concept of ISDB (Integrated Services Digital Broadcasting) for future broadcasting networks. The Communications White Paper in 1986 describes MPT's future prospects as follows: "As digitization of broadcasting networks progresses, it is desirable to construct an integrated system for various types of broadcasting services from the viewpoint of effective utilization of radio waves and also to offer various types of services more easily than is now the case. For this purpose, the use of multiplex signals is now being considered for various types of broadcasting services, such as radio broadcasting, TV broadcasting, teletext broadcasting, HDTV broadcasting and data broadcasting through broadband transmission networks based on the concept of ISDB."[3] In this vision, at the stage where broadband ISDN is realized and visual communication is available, ISDN and ISDB will be integrated into a "Universal and intelligent communication network". Technological convergence between broadcasting and telecommunications will be completed when the integration of ISDB and ISDN is achieved and the boundary between the two fades out as far as networks are concerned. Thus, convergence between broadcasting and telecommunications is also being propelled by the future vision of MPT.

3. Main Issues in Policy Making with Regard to Convergence of Broadcasting and Telecommunications

3.1 Recent Political Process

First, to briefly touch upon some political realities, all administrative competence to regulate both telecommunications and broadcasting in Japan is concentrated under the MPT. In the process of its policy-making, MPT refers to various councils established according to legislation and a number of ad-hoc commissions set up to deal with individual subjects. These organs serve as a forum for achieving consensus between MPT and other interested parties under MPT's initiative. Reports submitted by these organs give legitimacy to MPT's policies.

MPT and MITI (the Ministry of International Trade and Industry) became rivals to each other due to the convergence of telecommunications and computers. In spite of the need for integrated policy making, it is rare that coordination between the two ministries is made in a systematic manner at a cabinet level. In Japan an individual ministry tends to adhere to its own existing jurisdiction, sometimes in disregard of the realities of

[3] (In Japanese) MPT: White Paper for Communications, 1986, p. 796.

economy and society. This is called "TATEWARI GYOSEI (vertically divided administration)" and is often criticized by the public. The Diet makes virtually no contribution to the process of policy making in this field. Bills prepared by MPT and proposed by the government are endorsed by the conservative government party in advance and are hardly ever rejected in the Diet. Here one can expect some changes resulting from the goverment party's loss of majority in the upper house this summer. At any rate, interactive procedure between MPT and councils/commissions is the most important factor in the process of policy making.

There have been three recent political developments in response to convergence of broadcasting and telecommunications.

3.1.1 The 1988 Report of the Broadcasting Policy Commission and Revision of Related Laws

In April of 1987, the "Ministerial Advisory Commission on Broadcasting in the Age of New Media" (Broadcasting Policy Commission) submitted a report after two years of deliberation. According to the report, "it is our basic perception that the concept of broadcasting is changing. In regard to administrative regulation, the conventional dualistic approach towards broadcasting and non-broadcasting is no longer valid and regulations based on the closeness in nature of each medium to 'broadcasting' must be worked out. In other words, regulations in the light of characteristics unique to each medium must be explored in a relative perspective from the viewpoint of broadcasting."[4]

The Commission then proposed a policy of classifying media into those to which strict regulation should be applied and those in which regulation should be less strict based on the characteristics of each medium. On the other hand, the Commission took a very deliberate attitude towards the separation of hardware and software, saying that further consideration is needed in this field.

Agreeing to the suggestions made in this report, MPT proposed a partial revision of the Broadcast Law and the Radio Law. The revision was passed in the Diet in April of 1988 and implemented in October of that year. Its major features were the stipulation of a frequency allocation plan, partial relaxation of program-related regulations, and introduction of pay-TV service for private broadcasting, which would be provided via one channel of BS-3a scheduled to be launched in the summer of 1990. This revision reflects MPT's basic attitude of coping with the present situation and developments in

4 (In Japanese) Report of the Broadcasting Policy Commission - Perspectives of Broadcasting Policy -, 1987, p. 116.

the near future without altering the fundamental structure of the present broadcasting regime. Incorporation of pay-TV service into the framework of the Broadcast Law means that CUG service, which is ambiguous in nature and could be excluded from the category of broadcasting, will be also dealt with as one type of broadcasting service.

3.1.2 The 1988 Recommendation of the Telecommunications Council on the Review of the Telecommunications Business Law

The Telecommunications Business Law implemented on April 1, 1985 was subject to review within three years and thus became an important topic of consideration in late 1987. In the end, the Telecommunications Council concluded in its recommendation submitted on March 18, 1988 that "for the time being, the Council does not regard that the current framework of telecommunications systems requires any alteration and thus that any revision of the Telecommunications Business Law is necessary." MPT, in response to this recommendation, concluded that no revision would be made concerning the Law.

The Council, however, pointed out there is an urgent need to consider "a new rule for the 'boundary' service between communications and broadcasting." The Council's recommendation reflects the opinion of various concerned parties and organizations. KEIDANREN (the Federation of Economic Organizations), for example, argued for the establishment of "a new rule for the boundary service between communications and broadcasting", including a possible alteration of the present legislation. Contrary to most parties concerned who agreed to the view that it was too early to review the Telecommunications Business Law, MITI and its advisory organ insisted on the need for review. MITI's commission proposed a comprehensive reconsideration of all the laws and regulations related to telecommunications and broadcasting so as to make various types of boundary services possible in response to the convergence between communication media.

It should be noted here that a need for a new framework for boundary services was requested not from the side of broadcasting policy-making but from the side of telecommunications policy-making, and that it was industrial policy makers and representatives of non-media capitals who made the request.

3.1.3 The 1989 Interim Report of the Boundary Services Commission and Revision of Related Laws

In response to the recommendation made by the Telecommunications Council, MPT set up a Commission for Boundary Services between Communications and Broadcasting in August of 1988 mainly to discuss two issues during the period of one year. The first problem was concerned with services similar to broadcasting provided by CS, an urgent issue on the eve of the launch of private CS by JC-SAT and SCC. The second problem was whether NTT should be permitted to provide a cable television network. Both issues boil down to whether a Type I telecommunications carrier should be allowed to provide transmission networks for such services that are classified as broadcasting, that is, whether the telecommunications boundary should be expanded into broadcasting.

In regard to the first problem, concerning CS, immediately after the commission submitted an interim report in February of 1989, MPT proposed a bill to partially revise the Broadcast Law and the Radio Law for the purpose of realizing broadcast services via CS. The revision was approved in the Diet in June 1989 and will be implemented in October of that year. This issue will be discussed in detail in section 3.2.2.

The commission gave consideration to the second issue in the latter half of the term but was dissolved without submitting a further report because it was considered to be too early to propose any specific measures on that matter. Afterwards, a new ad-hoc commission was set up to take over the subject. This matter will be discussed in detail in section 3.2.3.

3.2 Major Issues and Policy Orientation

3.2.1 Telecommunications Services via Cable Television Networks

Presently, there is only one cable television operator that provides telecommunication services via its own network facilities. This is Lake City Cable Television (LCV) in Suwa City, Nagano Prefecture, which obtained additional permission to become a Type I carrier in 1986. The company allocates part of the frequency band of its coaxial cable to communications services by means of frequency division. This means that the same coaxial cable network facilities are operated under two licenses based on the Cable Television Law and the Telecommunications Business Law. LCV offers leased circuit services such as remote control of water resources, waterworks telemetry and a community medical system, and it also plans to construct a fiber optic loop in order to start LAN business.

The present subscription rate of cable television service is around 17% and most of the facilities are used for the purpose of improving reception conditions. However, recently a new type of cable television service called "urban-type CATV " was introduced into the market. Urban-type CATV is defined as having more than 10.000 service terminals, over five channels for original programs broadcasting, and a two way service function. It is characterized by multi-purpose services, in addition to retransmission of TV programs. MPT expects cable television to become a telecommunications media by equipping it with interactive functions. Cable television is considered as both broadcasting and telecommunications media and is expected to play an important role in the diversification of telecommunications networks. This means that a cable television operator will be expected to emerge as a competitor to conventional telecommunications carriers in local service.

MPT intends to expand the role of cable television, which is now fulfilling the minimum requirement of improving poor reception conditions, by allowing it to enter commercial business based on the concept of "Space Cable Network". This concept was introduced by the report of the Advisory Commission on Promotion of Space Cable Network in May 1988. In this concept, private CS and cable television facilities are linked to establish a new program distribution network. Thus a cable television operator, a program supplier and a CS operator will form a trinity to comprise a new industry. Before this can be realized, however, there are several problems which must be overcome. While existing cable television facilities, which are specialized in retransmission of TV programs, must be transformed by equipping them with multi-channels. Who will bear the cost of these improvements? To what extent will subscribers or installers share the cost? Also, potential for supplying programs must be increased. How is this possible? And after all, how will the legal responsibility for the contents of programs be shared between a cable television operator and a program supplier? The last problem will be discussed in the next section.

3.2.2 Broadcasting Service via "Telecommunication" Satellite

The core problem in the interim report of Boundary Service Commission was how to deal with direct, individual reception of services via CS. Among the services JC-SAT and SCC plan to offer are AVprogram delivery, music supply, on-the-spot broadcasting of events, business television, data distribution, program distribution, automobile dealer networks (auctions), education networks, religion networks, and SNG (Satellite News Gathering). Some of these are programs aimed at direct reception by unspecified people.

The Commission conducted interviews with telecommunications carriers (NTT, JC-SAT, SCC), broadcasters (NHK, NAB, JSB), the Japan Cable Television Association and the Association for Cable Television Programs Suppliers. Naturally they had different views on individual reception of boundary services. Broadcasters were against individual reception in principle and insisted that it should be classified as "broadcasting" service in case it was permitted. CS carriers argued that individual reception should be admitted as "telecommunications" service. Finally the Commission concluded that among boundary services using CS, information services that are received directly and simultaneously by unspecified persons, including payTV, would be categorized as "broadcasting" service, while the rest of the boundary services using CS would be classified as "telecommunications" service.

This classification will inevitably lead to the separation model. In the new system, "one who conducts broadcasting service will compile programs on his own responsibility and a CS carrier will control the broadcasting facility. The former will commission the latter to transmit broadcasting programs without any alteration of the contents to be transmitted and the latter will transmit programs as they are through its own networks."[5]

In the revised legislature which introduced this new system, a CS carrier is called "a commissioned broadcaster" and a program supplier is called "a commission broadcaster". A commissioned broadcaster is under obligation to obtain a radio station license for this purpose as well as to provide transponders. A commission broadcaster is required to obtain approval from MPT and is subject to the same regulations as conventional broadcasters, i.e. regulation of programs, exclusion of concentration, etc. The number of commission broadcasters is to be decided according to the frequency allocations plan.

Behind this legal solution, we find the administration's intention to maintain the current regulatory framework even after broadcasting service via CS is introduced and/or the separation model is adopted.

3.2.3 Cable Television Network Operated by NTT

Although NTT wishes to construct a cable television network, this has been opposed by MPT. In West Germany where the separation model has been adopted, Deutsche Bundespost is allowed to construct and operate a cable television network. This broadband distribution network has been expanded so as to supply a transmission

5 (In Japanese) Interim Report of the Boundary Area Service

network for commercial TV programs and to create new media markets. In West Germany, dualism of public and private broadcasters was made possible by linking this broadband distribution network with CS. Looking back at the history of Japan where dualism in broadcasters was introduced during the post-war Occupation, the development of cable television networks followed a different path. In Japan, cable television networks emerged as a means to overcome difficulty in receiving TV signals and also to transmit programs made by local communities. Thus cable television was a local facility to meet local needs. However, with the growth of "urban-type CATV", things began to change and NTT came to discover "economies of scale benefits" in constructing cable television networks.

NTT provides JCTV, a cable operator in Tokyo, with its circuit facilities and this is the only such example at present. JCTV provides hotels, apartment houses for foreign residents, and company and government offices with simultaneous retransmission of CNN programs from the U.S. as well as an English program of its own compilation using a video-transmission network leased from NTT. JCTV has permission from MPT to install a cable television broadcasting facility, which is actually a transmission network owned by NTT, meaning that this permission is fictious from a legal point of view. The usage charge for this network is fairly high because it is not a cable television network in its original sense but consists of circuits leased from NTT.

MPT has prevented NTT from constructing a cable television network and leasing it to cable operators since it wishes to: (1) maintain the local characteristic of cable television service; (2) to comply with the unification model; and (3) to diversify network types in order to overcome the vulnerability of the network system. NTT is being kept away from the cable television business for the purpose of nurturing cable television operators. However, it is not clearly stated how this principle of network diversification will be coordinated with ISDB or the broadband ISDN project in the long run. It is assumed either that local cable television networks will be maintained as telecommunications media in a transition period until broadband ISDN is achieved (even though this may end up with double investment on network facilities), or that the cable television "telecommunications" network will be developed into a regional broadband ISDN with consequent pluralization of integrated networks.

In relation to this, it should be noted that the NTT Law is scheduled to be reviewed within five years after its implementation, that is, by April 1990. The discussion in the Telecommunications Council, which is in charge of the review, is now centered on the matter of the proposed divestiture of NTT. Presently, it is impossible to predict what impact this review will have upon the problem of boundary services and broadband ISDN.

3.2.4 Broadcasting Services via NTT's Telephone Network

In August 1988, NTT started "Off-Talk Communications Service" after obtaining permission from MPT. In this service, voice and facsimile information is broadcasted through the subscriber's telephone network when it is idle. This idea of trying to increase revenue by using subscriber's telephone networks for purposes other than ordinary telephone calls, thus increasing the rate of network operation, reminds us of the case in the U.K. when videotex service was first introduced. However, NTT had to wait for a long time before it could realize this type of service because it was a quasi broadcasting service in the sense that information is transmitted simultaneously to a number of receivers.

While in providing off-talk service NTT is in charge of facilities and services between an information center and the subscriber's access line, including communications equipment for the service, an information center must be run by an organization other than NTT, for example, a local government, an agricultural or fishermen's cooperative. NTT has already started to offer the service in some rural areas in oder to meet the demand for an alternative service in those areas where wire broadcasting telephone facilities have become obsolete. In the meantime, a so-called "urban-type" off-talk service was initiated by a regional newspaper company in Hiroshima city in July 1989 to provide music and news. Depending upon the direction of future developments, off-talk communications services may become indefinitely close to broadcasting services which use telephone networks. Furthermore, assuming that narrowband ISDN will be extended to individual homes, off-talk communications services have potential to become widely diffused as a means for multi-address distribution of voice, facsimile and data, though not of visual information.

3.2.5 Broadcasting Satellites: Changes in Applications and Ownership

Originally DBS was developed by NHK, which is legally obligated to provide broadcasting programs throughout the nation and to solve reception problems relating to terrestrial program. The DBS was launched by Japanese rocket as part of the government's space development program. However, this situation is now changing as explained below.

NHK started a trial satellite broadcasting service with the capacity of two channels in December 1986 via BS-2b which was launched earlier in February. This service aimed at improving poor conditions for receiving TV signals through the retransmission of programs broadcasted by terrestrial waves. Then MPT revised the terms of licensing for BS-2 in June 1987 to expand its purposes to include satellite broadcasting services per se in addition to the improvement of conditions for TV reception. Subsequently NHK initiated 24-hour-a-day broadcasting of programs compiled for this service via DBS using one channel in July 1988. As a result of this, the number of households receiving satellite programs soon exceeded one million. In April 1989, NHK was permitted to switch its trial service into full-scale satellite broadcasting by MPT. From June, it started satellite broadcasting using two channels for 24-hours-a-day and also experimental broadcasting of HDTV for one hour a day. In August, NHK started to charge for its satellite broadcasting service. Recently NHK bought a satellite from General Electric Company of the U.S. as a backup for BS-2b, which is not sufficiently reliable, and plans to launch it by Arianne rocket in December 1989. The satellite, named BS-2x, is scheduled to begin operation in January 1990. Thus BS, with due consideration of HDTV broadcasting, is already on its way towards becoming one of the main broadcasting networks.

BS-3 is scheduled to be launched in the summer of 1990 and to start operation at the beginning of 1991. Among the three channels available, two will be allocated to NHK and one to JBS, a private broadcasting company. JBS plans to charge for its service in order not to influence terrestrial private broadcasters. As for BS-3b, which is scheduled to be launched in 1991, one transponder, which will be used exclusively for HDTV, will be possessed by Telecommunications Satellite Corporation of Japan and leased to HDTV program suppliers.

With regard to BS-4, MPT's inquiry commission submitted a report in February 1989, proposing that (1) all the eight channels allocated to Japan should be used; (2) coordination must be pursued with terrestrial broadcasting; (3) new types of broadcasting systems, such as HDTV, PCM sound broadcasting, facsimile broadcasting, and data broadcasting, should be introduced; (4) various types of business enterprises should be allowed to enter the market; and (5) a corporation to procure and control DBS and to lease its channels to broadcasters should be set up.

Establishment of a procurement corporation essentially means the introduction of the separation model. This reflects the fact that the immensity of the distribution network in satellite broadcasting has made it difficult to maintain the conventional ownership of broadcasting networks. Thus, we shall now have a kind of common carrier in the field of broadcasting networks.

Alltogether, it should be noted that the raison d'etre or core purpose of DBS has shifted gradually from solving problems in signal reception to diffusion of satellite broadcasting and finally to supply of transmission networks for HDTV. This, in turn, implies a shift from the public interest to the interest of the broadcasting enterprises and then to interests of an emerging industrial complex.

3.2.6 Telecommunications Services Using Broadcasting Transmission Networks

As two-way cable television networks are also capable of providing telecommunications service, broadcasting transmission networks based on radio waves can also function as distribution-type telecommunications media through multiplexion of terrestrial waves or DBS channels.

As mentioned above, the Boundary Service Commission dealt with the subject of inroads or advancement by telecommications into the domain of broadcasting. As for the issue of advancement in an opposite direction, i.e. from broadcasting to telecommunications, SCC, NHK and NAB among the interviewees confined themselves to hinting that it was necessary to consider the possibility of providing telecommunications services through broadcasting facilities (DBS and terrestrial waves).

However, it remains to be seen whether the multiplexed part of broadcasting networks possessed by a licensed broadcaster may belong to the broadcaster and whether it may be used for telecommunications business. In the area where a separation model is introduced, the transmission network will no longer be classified as a broadcasting network but will have the characteristics of a telecommunications network of a common carrier, consequently leaving no room for uncertainty.

4. Theoretical Considerations

4.1 Functional Concepts of Broadcasting and Telecommunications

What are the typical concepts of telecommunications and broadcasting? They tend to be deduced from definitions and terms used in the current legal system. One cannot deny the fact that these concepts, which are vital in any analysis of the convergence of telecommunications and broadcasting, are strongly influenced by legal interpretation. As legal considerations were already given in section 2.2, here we will try to conceptualize broadcasting and telecommunications from the viewpoint of their functional difference.

Both telecommunications and broadcasting are social systems for distribution and exchange of goods/resources, i.e. information or information services. We can extract the two fundamental factors governing this system. They are:

(i) the mode in which information goods are supplied,

(ii) the scope of the supply of information goods
and/or
the qualifications for access to them.

(i) is concerned with the purpose or intention of information transmission and the pattern in addressing on the side of the sender. It can be further divided into distribution and exchange. (ii) is the scope of supplying information from the viewpoint of the sender and the qualifications for access to information from the viewpoint of the receiver. Basically where qualifications are needed, access to information is closed and where qualifications are not needed, access is open. Figure 1 comparatively illustrates the concepts of broadcasting and telecommunications according to the required qualifications for access to information (X-axis) and the purpose of information transmission (Y-axis).

Figure 1: Functional concepts of broadcasting and telecommunications

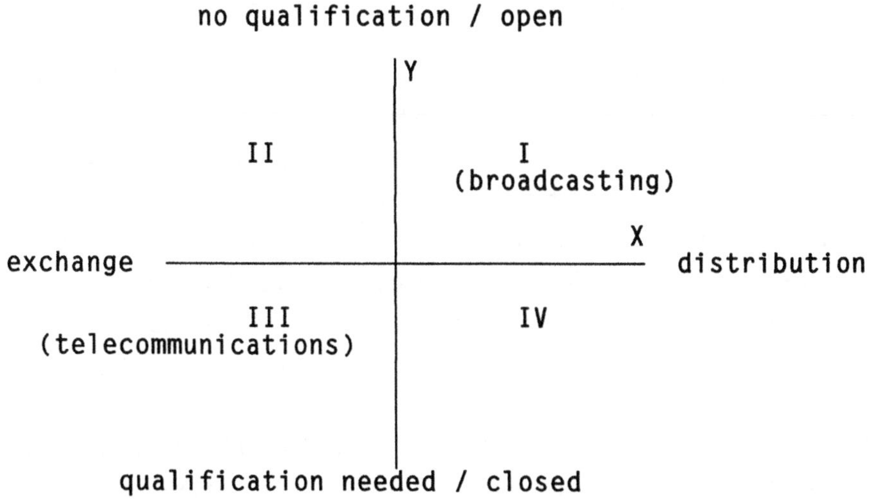

X-axis = purpose of transmission
Y-axis = access qualification

Quadrant I indicates the domain of broadcasting functions. Broadcasting is a system which aims at the distribution of information and requires no qualifications for access to information. Quadrant III corresponds to the domain of telecommunications functions and is defined as a system which aims at exchange of information and requires qualifications for access to information.

As a corollary, broadcasting and telecommunications can be characterized as follows from the viewpoint of network application and to whom information belongs. Quadrant I is characterized by information sharing as well as shared use (not co-ownership) of networks by the public. There, public information is distributed. Networks are divided into radio and wire depending on network technology. Quadrant III, on the other hand, is characterized by closed information and exclusive usage of networks. Information is exchanged between closed parties. Networks are divided into public and private. Public networks are available to any one and become exclusive only when they are used by qualified parties. Private networks are available to particular users who have leased them from a carrier. Regulatory measures and rules of control towards transmission network business vary according to these differences in scope of information supply and network application.

However, in the process of convergence between broadcasting and telecommunications, the traditional correspondence of one transmission network to one type of service has broken down. At this stage, it is no longer possible to define either broadcasting or telecommunications as a system composed of networks and services. Now they must be looked at in terms of the individual services each network provides.

As a result of progress in information and communication technologies and diversification of users' needs, a new sphere of service has emerged between broadcasting and telecommunications services. It is a sphere in which information is shared among qualified members only, as indicated in quadrant IV of the Figure 1. It is the domain of CUG, including the conventional multi-address communications service. In this area restrictive public/closed information is distributed. Two groups are identified here according to the type of access qualifications. These are a multi-address transmission group and a paid service group. In the former case, regardless of whether the user pays for the service or not, restriction of access is not lifted and information is closed. In the latter case, restriction of access is lifted and information is disclosed on condition that the user pays for the service. Furthermore, in a paid service system where economic factors play an essential role, the scope of distribution, in other words, the optimum size of the market, has a significant meaning.

Now the problem must be considered in the framework of convergence between broadcasting, telecommunications, and computers. In this framework, types of communications are divided into value-added communications and transparent communications depending upon whether or not information processing is performed.

In Figure 2, the Z-axis representing computer functions, is incorporated into the X and Y-axes. Each of the eight spheres created by the three axes corresponds to one service area with a different function.

Figure 2: Three-axes-concepts

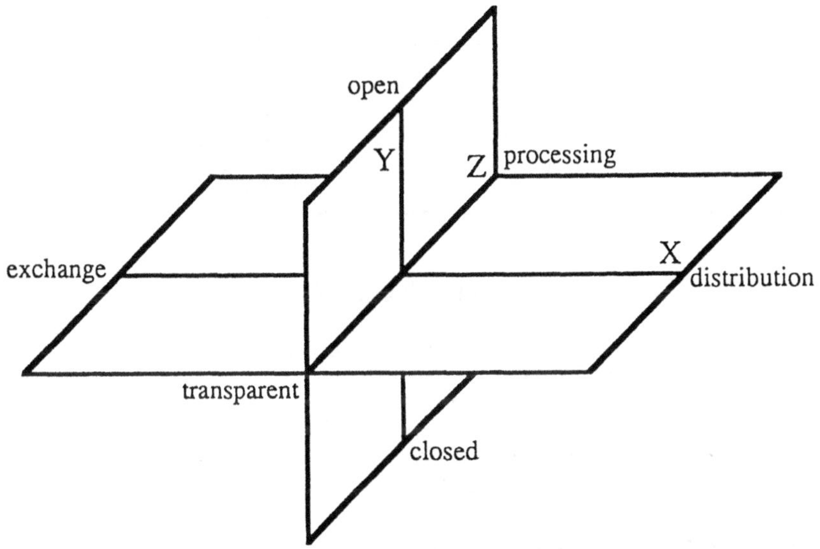

X-axis = mode of supplying information goods: distribution or exchange
Y-axis = access qualification for information goods: open or closed
Z-axis = information processing function: processing or transparent

Figure 3 shows functional concepts composed of octants, which will be named as follows: I-classical broadcasting service, II-(potential sphere), III-classical telecommunications service, IV-closed users group delivery service, V-(potential sphere), VI-information reference service (data base service), VII-value-added service, VIII-(potential sphere). They are at various stages of development. Although the service spheres of II, V and VIII are only embryonic at present, they have a certain potential for future development from the theoretical point of view.

Figure 3: Functional concepts composed of octants

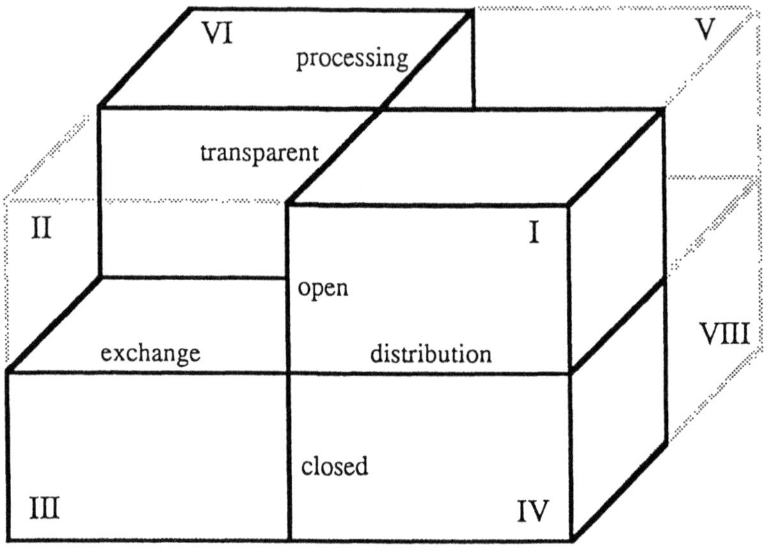

I	classical broadcasting services
II	(potential sphere)
III	classical telecommunications services
IV	CUG services
V	(potential sphere)
VI	information reference services (data base services)
VII	value added services
VIII	(potential sphere)

4.2. Crossover of Networks and Services

Convergence of broadcasting and telecommunications can be summarized as follows based on the functional concepts explained above. The convergence has produced two kinds of transformation. First, the fixed relationship between one transmission network and one type of service has been destroyed as the networks are equipped with more functions than before so that one transmission network can provide various types of services. As a result of this, telecommunications networks are now capable of providing broadcasting and CUG-type services on the one hand, and broadcasting networks are capable of providing telecommunications and CUG-type services on the other hand. Examples of the former include broadcasting services by CS, off-talk communications service, and business television. Examples of the latter are telecommunications services via two-way cable television networks, pay-TV, and paging services via FM multiplex channels. Consequently, one transmission network facility may be subject to more than one type of a license and regulations depend upon the types of services it provides.

4.3 Crossover of Business Domains and Marketplace

The relationship among networks, the business domain of an enterprise and the marketplace, which used to be fixed in the past, has also been transformed as a result of the crossover of networks and services. Subsequently, an enterprise expands its business domain into other marketplaces. This is the second type of transformation. There are also two directions in this process. Telecommunications carriers enter broadcasting and the CUG service market on the one hand, and broadcasters enter telecommunications and the CUG service market on the other hand. Examples of these include cable television operators with a carrier's license for Type I telecommunications, NTT's construction of a cable television network (although this has not been realized yet), and CS carriers which are also commissioned broadcasters.

5. Conclusion

5.1. Technological Convergence and its Industrial Consequences

5.1.1 Regarding Industrial Organizations

Technological convergence leads to changes in service production organizations. Technological convergence of telecommunications and computers brought about structural changes in telecommunications, i.e. diversification of telecommunications organizations and markets. The telecommunications industry has been divided into the field of carriers and that of service providers. Various types of enterprises have emerged in each field, although it can not be denied that there is still one dominant enterprise at large.

In such a liberalized environment, the technological convergence of broadcasting and telecommunications is giving rise to structural changes in broadcasting, which also take the form of diversification of organizations and markets. The broadcasting business consists of three component parts: the construction and operation of transmission networks; the production and compilation of programs; and the means to procure financial resources. Although these three parts have been conventionally controlled by one organization unit and business has been conducted in an integrated manner, they are gradually being disjointed as a result of the technological convergence and diversification within each part. Transmission networks have diversified into terrestrial waves, cable television networks, satellite transmission, and multiplexion of each network. The programs market are segmented according to audience preference. Financial sources are diversified into reception fees, advertisement, and charges for pay-TV, and the reception fees itself are further divided into fees for terrestrial programs and those for DBS programs. Diversification of transmission networks leads to changes in ownership of network facilities. Segmentation of the programs market reflects as well as promotes the diversification of the audience. Diversification of financial sources is considered to reflect diversification of production forms for broadcasting services.

What should be noted here is that as a result of diversification in various parts of the broadcasting business, diversified elements (sub-parts) are combined in various ways. In this way, the production of various small-scale services as well as phenomenon called "media-mix" and "multi-outlet" have come into existence in the world of mass media including broadcasting. In other words, technological convergence has caused the disintegration of various parts of the broadcasting industry into sub-parts, after which the discretionary combination of various sub-parts has been realized.

5.1.2 Regarding Industrial Structure

Technological convergence also brings about changes in the relationship among various industrial sectors. With the convergence of telecommunications and computers, such industries as computer and terminal equipment manufacturing, the automobile and transport infrastructure industries, and also electric utilities, have entered the domain of telecommunications. For example, by new satellite common carriers, JC-SAT has a capital composition of C.Itoh & Co. Ltd. (40%); Mitsui & Co., Ltd. (30%); and Hughes Communications Co. (30%). SCC has a capital composition of Mitsubishi Corporation (26%); Mitsubishi Electric Corporation (20%); and Mitsubishi Heavy Industries Ltd. (10%).

With the technological convergence of broadcasting and telecommunications, non-media capital has come into broadcasting industry. This was made possible by organizational changes in the broadcasting industry. The disintegration of the broadcasting industry into various components as mentioned above prepared the conditions for capital other than in mass media to be engaged in network operation and programs supply. What should be noted here is the activities of the general trading companies. In Japan there are several groups of affiliated companies which tend to dominate the Japanese economy. Each group usually consists of a main bank, a general trading company and affiliated manufacturing companies. The general trading company plays the role of a coordinator. Now all of the major trading companies have a subsidiary to supply programs, which are sometimes imported from abroad, via cable television network. Some of the department stores are producing audio-visual programs to give themselves some distinctive cultural trait. They obtain a distribution network for their programs by means of broadcasting via CS. The trading campanies have invested in CS network operation and programs supply, respectively, and have established a position to control both of them.

In Japan, capital concentration within the mass media reached a certain level by the middle of the 1970s and now the process of capital concentration has entered a new stage on a larger scale including capitals of other industries that joined this process recently. It is by no means a coincidence that HDTV, which requires an enormous amount of capital for programs production, is going to be put into actual use in parallel with this process of capital concentration.

5.2 Technological Convergence and its Regulatory Consequences

Institutional conflicts that have been brought about by technological convergence require coordination among the various institutions concerned. The convergence of telecommunications and computers brought about institutional reform towards liberalization. This was, in other words, the result of the fact that the principle of competition in the computer industry, i.e. the manufacturing industry was introduced into the field of telecommunications as its economic potential increased.

The convergence of telecommunications and broadcasting, however, has not yet brought about a fundamental reform of the broadcasting system in Japan. Technological convergence entailed regulatory diversification according to the characteristics of the media. Thus the unification model, the fundamental principle of the broadcasting industry in Japan, was partially thrown away in the sector of private CS and replaced by the separation model. This was because the separation model was more suitable from the view point of competition and economic efficiency as explained in the previous section. The separation model was introduced into CS service because its economic potential increased so much that it became impossible to cope without a new organizational structure.

However, in spite of the introduction of the separation model, the status quo of administrative regulation under the unification model of networks and program supply has been maintained. It is inevitable that the legitimacy of MPT's regulation, which springs from technological licensing of transmission networks for broadcasting, will be questioned under the separation model, where responsibility for programs and that for the networks are separated, and therefore regulations for each of them must be arranged differently.

The convergence of broadcasting and telecommunications is considered to be the process where developments of transmission technology and changes of its applications and ownership forces a transformation of the transmission order as a whole. This requires a new type of relationship among the four factors: the network operators, network users, regulatory body and the audience. None of them should be isolated in this process of transformation. Although it is necessary for the regulatory body to arrange a new type of organization and regulatory methods based on new concepts, taking up this issue has been avoided in the process of Japanese policy making because the established interests of various ministries are concerned.

The current regulatory framework of broadcasting was formed in the age of the electromagnetic wave and has been maintained by expanding the concept of broadcasting. However, during the age of broadband ISDN, this framework will be no longer relevant and a new concept of broadcasting and a new regulatory framework based on it will be very desirable.

5.3 Responsibilities for Cultural and Social Values

The convergence of broadcasting and telecommunications has been promoted and transmission networks have been expanded with increasing complexity by techno-economic factors in Japan. Under such circumstances, much greater responsibility is given to policy making than ever before. In a democratic and pluralistic society, it is indispensable to redefine and consolidate institutional values unique to broadcasting and telecommunications.

In the field of broadcasting, the cultural values (or roles) that have been maintained so far are gradually eroded and disturbed over a wide range. However, this does not mean that the institutional significance of broadcasting for a democratic society has disappeared. Rather, a system for creating some kind of public, universal, or basic values is needed in a diversified and complex society. While administrative policies tend these days to be made on the basis of economic potentiality or the possibility of innovation, restructuring the institutional framework from a social and political standpoint is highly desirable.

In the field of telecommunications, it is the goal of our society to reconfirm and guarantee protection of the rule of "secrecy of communications". Along with the technological convergence, the distinction between communications which should be kept from disclosure and those which need not be kept secret is blurred. In other words, the boundary of the domain where the rule of "protection of secrecy" must be observed becomes ambiguous. Therefore, there arises a need to reconsider the principle of "protection of secrecy" as well as to reconfirm the rule given technological developments and new types of services. However, in addition to the protection of "secrecy of communications," which is a passive aspect, we also have a social and political responsibility to establish a system to guarantee human rights in relation to the networks including the protection of data concerning individuals, which is an active aspect, as transmission networks grow to be increasingly enormous and complex.

References

Broadcasting Policy Commission (1987), Report of the Broadcasting Policy Commission - Perspectives of Broadcasting Policy - (In Japanese)

Commentary for the Telecommunications Business Law (1987), (In Japanese)

Ministry of Posts and Telecommunications (MPT) (1986), White Paper for Communications (In Japanese)

13

Telecommunication and Audiovisual Legal Regimes in France

Bertrand Delcros[1]

Chargeurs SA
Paris

Jean-Pierre Chamoux[1]

Association
Droit & Informatique
Paris

1. Introduction

Between 1978 and 1980, French courts have decided on several cases dealing with private radio broadcasting in violation of the broadcast monopoly. At the time, there were no penal sanctions in the broadcast statutes. But some jurisdictions thought they may assimilate broadcast penalties with the existing penalties included in the télecommunication code: this was the start-up to bridge the gap between audiovisual and telecommunications legal regimes.

The 1982 Audiovisual Law has stated the freedom of audiovisual communication. French audiovisual law was ready to start. But this raised several questions:

- How should it relate to telecommunication law?

---

1 Based on B. Delcros draft paper for EUROCPR 89, this article has been completed and translated into English by J.P. Chamoux.

- How should it be inserted into the existing law?

- Should it be considered as an autonomous legal regime?

The licensing system, which is by now a common principle for broadcast and for telecommunications as well, can be considered as a witness of this evolving rule of law. One might start in thinking that Audiovisual is bound to the general public (or to shares of the general public) while Telecommunications is characterized as a private transaction. Most of the existing rules are taking that for granted. We shall see, however, that such a distinction solves only some of the problems at stake, and that more clarification has to come in the future.

2. Licensing

2.1 The Responsible Body for Licensing

2.1.1 The Body for Audiovisual and Broadcast

The French statutes have now clearly given the licensing power to public and non governmental bodies respectively:

- Haute Autorité de la Communication Audiovisuelle (from 1982 to 1986) - HACA,

- Commission Nationale de la Communication et des Libertés (from 1986 to 1989) - CNCL,

- Conseil Supérieur de l'Audiovisuel (since January 1989) - CSA.

Hence, the public broadcasting monopoly, established in 1945 in favour of the State, has vanished since 1982. However, in the meantime, the political interference with audiovisual media which was the rationale for the State monopoly after the war, has not been really eradicated. Changes in political leadership have lead to several modifications in the audiovisual law over the years, witnessing the permanent temptation of interference between government and these media.

As an example we refer to the 1982 audiovisual law stating that "Communication is free" in article 1 of the statute. The law formed an "independent administrative authority" (HACA) which pretended to be free from government influence. But a real independence is not easy to achieve! The conservative majority elected April 1986 decided to modify the law and changed this body into another one - CNCL - with a different nomination scheme and enlarged powers. The socialist victory at the legislative and presidential elections in June 1988 lead in turn to a new change of the name of this body - now nominated CSA - and a few minor adjustments of its powers.

This chronicle instability in the poeple, in the staff, in the powers does not facilitate a long term policy. However, one thing has been stabilized at least: the licensing power for broadcast operations is now out of direct government hands. And, over the years, this non governmental body has gained more power: HACA was only able to license local radios and TV stations (radius 30 km for radio stations, radius 60 km for TV stations). CNCL and CSA, respectively, have the ability to license all communication services whether national or regional, but for the state stations which are regulated differently: as an example, chairmen of public stations are nominated by CSA, but public service obligations are not stated by CSA.

2.1.2 The Body for Telecommunications

The state privilege there has not been fundamentally reconsidered until 1986. The 30th September 1986 Communication Law has brought about for the first time a serious legal breach in the state telecommunications monopoly. Article 10 of that law did treat differently private dedicated networks and networks opened to third party traffic. The basic idea was to let CNCL license the first category of networks (basically: dedicated infrastructures, whether wired or radio), and maintain the original licensing power within the government administration (i.e.: the PTT ministry) for a while. The objective was to test this over a period of two years (1986/1988) and to prepare, during this period, a more deregulated environment which would have been regulated by CNCL according to a competitive policy to be stated in a special law by the end of 1987. A draft was released in August 1987, but was never enacted prior to the change in Government in June 1988.

The present Government, acting since then, has decided to revert this policy and to bring all licensing power back into the government, and under the authority of the Minister for Post and Telecommunications. A statutory decision should be taken early 1990 to confirm that new policy line. Meanwhile the CSA is only exerting a mock responsability in that field, most of the daily decisions being taken within the administration itself.

2.2 Licensing Procedures

2.2.1 Procedures for Audiovisual and Broadcast

The rule is a public call for proposals, prior to any licensing for new frequencies or new regions. But, as we said above, political interaction is somewhat hard to avoid in this process! The licensing of the 5th and 6th National TV Channels were a spectacular demonstration of these political mishaps.

By the end of 1985, the former socialist government decided to license these TV Channels to two consortia formed very quickly. The terms and conditions of these "public service concessions" were not cristal clear. But they were in line with the enforced 1982 audiovisual law. The new conservative government formed after the Spring 1986 elections resiliated those licences immediately. The CNCL, formed in October 1986 according to the new statutory rules, did issue a call for proposal for reallocating these two national channels.

This 'call for proposal' procedure is of old practice in some fields like public works or public procurements in France. It is considered as a guarantee for some transparency and a fair choice. Candidates file an application with much data concerning their plans, finances, programs and affiliations. Public hearings can be held prior to the final selection. As a matter of fact, such public hearings are now compulsory for CSA licensing. Then, the regulatory body makes a notice of his decision and states usually the rationale for its choice.

In the past 7 years, this public and rather heavy procedure has been finally agreed upon for broadcast licences, and it is the rule for any new licence issued by the CSA.

2.2.2 Procedures for Telecommunications

The procedure in this field has been much lighter. The PTT Code states that the minister in charge of telecommunications can license infrastructures or services, or both; the law does not imply any formality or any obligation for the licensing authority to issue a public call for proposal. This can be understood for there is little risk that a new operator or a new network may significantly modify the diversity of beliefs or behaviours in a democratic open society.

However, there has been some concern that allocating new frequencies or opening new opportunities to private operators in telecommunications might breach another economic principle: fair competition and equal economic opportunity for the potential contenders.

On that ground, it has become natural to issue a formal consultation for those major licences like a cellular operator or trunking radio networks. In those cases, the licensing authority (The Ministry) has published a call for proposal and followed a similar procedure to the one quoted above for TV networks. The CNCL and CSA, respectively, were also able to organize hearings and make their voice about the selection process. But for many minor licences (paging, VAN'S, teleconferences) the procedure was much simpler, the new ministerial decision being made as a purely administrative licence.

The PTT department has been reorganized in September 1986 to integrate this new responsability into the administration: first, a special task force was formed (La Mission à la Règlementation) which was notably responsible for handling licences, allocating frequencies, regulating VAN's, etc. This body was the administrative counterpart to the CNCL/CSA Services. By June 1989, this task force was consolidated into a directorate which may receive more competence in telecommunications standardisation and licences in the next year or so.

2.2.3 In Summary

Telecommunications and Audiovisual operations are both bound to licence agreements in France. Whether issued by an administration in the government (i.e.: telecoms commercial operations) or by a non-governmental authority (i.e.: audiovisual commercial operations) those licences are considered as administrative decisions. They are ruled by administrative public law, and eventually disputed before the administrative courts and the Conseil d'Etat. But, as it is often the case for administrative decisions, the courts' control is restricted rather to formal procedure than to equity principles, under the French administration tradition. Yet this might very well evolve in the future, under the influence of the fair competition rules of the Treaty of Rome which will cover more in the fields of what were traditionnally the state prerogatives.

3. Licences

3.1 Licences for Physical Networks

Whatever the purpose - audiovisual or telecommunication in general - emission and reception are fundamental parts of a communication process. Reception is considered as a freedom in France. Emission is, on the contrary, constrained by a specific regulation and by law. Terms and conditions of transmission and broadcasting licensees differ according to the purpose of the emission.

3.1.1 *Network Licences for Audiovisual*

The broadcast monopoly has been progressively abolished over the years and the successive changes in the broadcast regulatory scheme. The law enacted on the 29th July 1982 on the eve of the first socialist government since the gaullist era, has abolished the monopoly given to TDF (Télédiffusion de France) since 1974 when that public authority was founded. This law was establishing a distinction between the use of a given frequency to broadcast and installing a broadcasting facility (i.e.: the infrastructure versus the service). The government was granting the licence to use a frequency, was licensing the facility, whereas "publishing" was, in broadcast mode, either licenced by the government or by the High Authority (HACA).

The law passed on September 30, 1986, has modified this system: the new regulatory body, CNCL, was allocating the frequency, the service and the facilities as well. The CNCL follower - CSA - today in power has kept the same powers to license any facility and service in these media.

Hence, the audiovisual regulator, established outside of the government administration, has full powers to regulate and allocate those parts of the spectrum allocated to radio, TV, and the like. This body controls, monitors and licenses FM band frequencies, TV channels, and has the duty to guarantee the efficient and comfortable reception of broadcast signals on the French territory. On wireless, this is not an easy task in large cities, for quite a number of stations are competing for audience. Several hundreds have been licensed with a changing pattern in the allocation process and their obligations. Networking has started already, and lobbying by these media is far from inefficient.

There is an obvious link between the licence to broadcast and the right to build up a base station. For national TV networks, CNCL first, followed by CSA have given way for the 5th and 6th TV channel to extend their coverage to a nearly national one.

More recently, the law has taken a step to qualify the use of broadcast frequency. There has been quite a debate on whether or not frequencies can be appropriated (i.e. a real asset). Diverging opinions have been published and case law has been reluctant to make this point clear. The last amendment to the communication law, on the 19th of January 1989, has made a new point. The main doctrinal positions were the following, prior to this enactment:

- one view is to consider the radio spectrum as "res nullius", giving some backing to the principle "first in, first served" which, by the way, has been accepted in history at the international level, given the way ITU frequency mapping was build over the last century;

- the alternate view was to consider the spectrum as "res communis" (i.e.: a public good) from which follows the fact that neither private appropriation is feasible, nor a public one, frequency access being arranged according to social or economic priorities organised by the state authority on non-proprietary grounds;

- a third view was to integrate the spectrum into a public domain (i.e. a propriety of the State or a "collective" good). From this flows the principle that licensing a frequency to a private person gives this person a licence to run a public property as a leaser of this public property. This qualifies the licensee in a similar way as if he was given a lease on a public park, in the side walks or on the beach.

The 1989 law has chosen the third hypothesis, which is an original choice, quite different in practice from the other two doctrines. From this several consequences may be derived that may alter, in the future, the process to allocate frequencies and the way businessmen behave with respect to these licences. It will be particularly interesting to see how this will match the EC integration for frequency management and transborder communication practice.

3.1.2 Network Licences for Telecommunications

There has been a series of attempts to review the aging telecommunication privileges of the PTT administration. The 1923 basic review of the telecommunication code - basis of all practice since then in France - had established a biased advantage in favour of this administration, although a licensing system did exist and was effectively applied in many practical situations: the railways, power and gaz companies, pipelines, some industries, the metro, firemen, and many other users were granted licences to lay down and operate private facilities and base stations outside of the PTT monopoly. However, those networks were in general self supported, and were unable to interconnect with the public switched networks and barely accepting third party traffic, but in very limited and restricted cases.

The more competitive regime promised by the 1986 legislation has not been enacted: a draft law has been established and circulated in August 1987, but it has never been introduced in the Chambers. After a period of uncertainty, the present government has announced a new draft legislation in the telecommunication sector, expected to be released early spring 1990.

There is a probability that this draft may introduce a limited monopoly constrained by the ten positions embedded into the EEC Green Paper for a telecommunication single market. But no major release of the passed tradition is expected, given the general socio-political environment in France and in Europe at large.

TV Cables might be an interesting illustration of the problems at stake for the physical networks, although the cable TV has been only dedicated to audiovisual programs with the existing operations licensed in the French Territory.

A vast national cable network was planned at the end of 1982, known under the acronym "Plan - Cable". This was supposed to finance a national coverage of the French territory, all in digital fibers, with an exclusive licence given to the French telecom administration to finance and realize that plan. Along the years, it has become obvious that this plan was a mistake: it has not been completed, although some 52 cities have signed a contract with France Telecom to comply with this scheme. Only a few of these were effectively all fiber. Most of the recent implementation have a mixed technology with some fiber back bones and a copper local loop.

The September 1986 communication law has learnt from that failure : the monopoly on the infrastructure has been abolished (art. 34 of the law). Local authorities are now able to license a company for TV cable installations, the use being specifically limited to audiovisual services-as stated in the law.

Hence, there is no visible infringement to the France Telecom privilege to establish public switched multipurpose telecom networks in this legal move. Nonetheless, it has opened a breach into the monopolistic tradition, as well as it was the case with the cellular telephone licences granted in December 1987 to a private consortium led by a water supply and utility company. This network is now commercially operating since March 1989 and covered some 20.000 subscriptions early 1990, with a fast growing demand: 10.000 enrolled in the last bimestre of 1989, mainly in urban areas around Paris, Lyon, Marseille, Lille, Nice...

The issues of multiple networks are thus raised in practice, although of limited scope, given the advantages of France Telecom networks and services, both modern and efficient as compared to most of its foreign counterparts in other countries of Europe. It remains to be seen whether a multiple network situation is a workable solution on the long run or not. Many towns and cities have given way to private and public companies to lay down TV Cables in their neighbourhood. The commercial return is still uncertain, but it is now up to the point where no return is possible for the next years.

3.2 Service Licences

3.2.1 Licences for Audiovisual Services

The licence structure and licensing regime are governed by a sense of pluralism: most of the rules and obligations given to the licensee tend to maintain, as well as possible, the plurality of opinions and beliefs embedded in the Constitution (freedom of thought etc...). This is of course of great significance, but it remains to be seen whether this may create the conditions for a successful business or not. Two cases will illustrate the stakes.

- The Constitutional Council has required a very stringent regime to control media concentration and ensure fair competition in this domain. The resulting rules are very complex and almost impossible to describe in plain terms. Just note that no one can, in principle, hold more than 25 % of a TV national channel. No majority control can be exercised on the licensed companies which makes it harder to run those businesses in practical circumstances. A clear demonstration of this difficulty was given with a tentative take-over on the French Channel 5 which failed early 1990.

- On grounds of limited frequency resources for national networks, the legislator has introduced a series of "general interest" obligations to be now born by the TV Channel licences, inspired by and similar to those obligations born by the public broadcasting authorities (A2/FR3): these refer to the amount and frequency of advertising, to movies programmed, to the support given to French (or European) producers, to copyright, etc...

Those obligations are serious constraints for a privately operated business. As a matter of fact, the private operators tend to play at the edge of those rules (TF1, La 5, M6). And the most successfull of these have gained an atypically advantageous regime under a concession specially granted and designed for this pay channel (CANAL +).

The audiovisual services are developping a quasi-editorial responsability, which resembles the system of editoral licences under the royal privilege in ancient France.

Comparing this constrained regime with the one of telecommunications services is quite illustrative for the differences which may, however, not be sustainable in the long run in an open and competitive Europe.

3.2.2 Licences for Telecommunications Services

The principle stated in the P+T Code (1923 law and art. L33 of the Code) is a general licence for all telecommunications services. As we have said earlier, third party traffic and commercial operations have been under strict control along history, the telecommunications administration maintaining a strict control over these services and providing most of them.

A few exceptions are worth quoting however: third party traffic and commercial service was the business for TDF from 1974 to 1988 when this public company was taken over by the financial holding company COGECOM, controlled by the French administration. These commercial ventures continue today, but they are strictly limited in scope and clients: TDF is serving the audiovisual operators, public and private.

Paging and trunked private telephone networks have been licensed to private companies, but on a limited geographical area and/or to companies controlled by France Telecom - Cellular radio telephone is a competitive market now, with France Telecom system and the new "ligne SFR" license quoted above. This was designed to introduce real competition in this field where France has been lagging behind the leading European countries for six years (Nordic countries and UK).

A third case is brought about by "value added services" operated on either the videotex standards or on leased lines networks. From the origin of TELETEL, the French videotex has been operated somewhat as a free market: besides the electronic directory - which by the way accounts for half of the whole TELETEL traffic figures - most services accessible on the Minitel are run by companies and with a free entry: there is no licensing needed to start a service or a data bank. But, after some trials and many errors, the "electronic publishing houses" have been asked to follow the general publishing rules, and a registration scheme has been established with the audiovisual regulator (CNCL first, CSA now).

The operation of leased lines based value added services - companies like GSI, AXONE, GEISCO, SLIGOS, CCMC and hundreds of others - has been considered in France as worth a special attention, like in the UK or in Germany lately. Started in a no man's land in the 1960's and 1970's, those services have grown big in the 1980's and were "tolerated" by the administration on unstable ground until September 1987. A decree has then released most of their limitations to third party service and interconnection of their clients on a national network, provided three conditions are fulfilled:

- the operation should register with the P+T regulatory administration, which has been separated from France Telecom in 1986 but still reports to the same Minister in the government;

- no commercial voice resale is permitted for the time being;

- large networks should offer open standards and internationnally admitted protocols as an alternative to the proprietary protocols they migh wish to market to their clients.

For the rest, telecommunication services are ruled by a general obligation to protect the privacy of person-to-person or business-to-business or business-to-person conversations and transactions. A few exceptions are tolerated to wiretap or to monitor the services under a court order or a subpoena. Most cases relate to smuggling, drugs, criminal offenses or espionage. And they are not a daily practice in modern times.

4. Concluding Notes

During the last decade, the broadcasting monopoly has vanished. A private sector has seized these new opportunities in radio and television services. Meanwhile, the general telephone network was restored to a high level of efficiency, in contrast with the degraded French network of the 60's. Taking this into account, the French public and the French Governments active during this period were not really motivated to change drastically the way telecommunications were run in France: minor adjustments were made, notably to raise new opportunities in Value Added Data Networks and radio communications.

As a consequence, the telephone monopoly is still active in France. Hence, after a period (1986/1987) where a converging regime for audiovisual and telecommunication was expected, we are now bound again to a diverging set of rules: diverging in principles, diverging in practical terms and conditions, diverging with respect to the regulatory body in charge of these rules.

French Telecommunications and Audiovisual communications are ruled by a mixture of public law (frequency allocation, licensing), private law (corporate practice) and penal law (torts and editorial liability). The reference to telecommunications rules are vanishing from the audiovisual rules. But, the other way round, more often than not, telecommunication operators are carrying audiovisual signals on their networks (Telecom 1 for la 5 Cable TV in principle) and videotex services have to register with the audiovisual regulator!

Are we leaving a mythic period for convergence, or are we living a real separation of those two fields? Opinions may differ, according to the perspective. At first sight, the old principle "Singula generalibus derogant" seems out moded in France. Shall it come back into force soon? Mobile radio and Cable TV are the two fields at stake for the coming years, and both may force the French legislator into a new realignment of legal practice in telecommunications and audiovisual operations.

Comment Part V

Nicholas Garnham

Centre for Communication and Information Studies, Polytechnic of Central London London

All three papers deal, from their different perspectives, with the regulatory problems raised as nations plot their different courses towards a common broadband environment for broadcasting and telecommunications.

We are just leaving a simple dichotomous world in which the distinction between broadcasting and telecommunications was clear cut. Broadcasting offered universally distributed point-to-multipoint services free at the point of consumption transmitted by over-air terrestrially based radio networks. Telecommunications provided transparent point-to-point voice and data communication links between individual subscriber terminals charged on a usage basis. In broadcasting content was heavily regulated. In telecommunications content was unregulated.

We are now moving down a migratory path towards a broadband environment within which it will be possible to deliver the whole range of communication and information services to home or office via a single broadband network. As we begin this migration the clear distinction between broadcasting and telecommunications starts to blur. How should we conceptualize and regulate (if regulate at all) cable TV networks, subscription TV, closed user group data broadcasting services, mixed use communication satellites and so on.

Furthermore should we assume a unified broadband infrastructure as the desirable goal of policy and so plot a programme of regulatory change to optimise movement towards that goal or, on the contrary, should we remain technologically neutral and so shape regulation as to allow the market to dictate the most efficient, perhaps fragmented, solution.

The Main Issues of the Papers

All three papers, in their different ways, deal with these problems. Delcros deals with a specific national legal problem, namely how France has, in recent years, approached the problem of the relation between broadcasting and telecommunication law. I will not refer to this useful paper further except to note that similar problems are being faced within other national legal systems and traditions. One of the issues we might discuss here today is the extent to which the legal problems and solutions raised by convergence are similar from country to country or whether on the contrary distinct national traditions and legal systems may cause countries' development paths to diverge in significant ways. This is an issue of particular importance within the European Community as it attempts to harmonise its regulatory arrangements and markets.

Hanada and Lera's papers take a wider view of the problems raised by convergence. For Lera even if a converged broadband scenario offers clear technico-economic advantages based upon avoidance of the duplication of infrastructure investment and the saving of scarce spectrum, such convergence is likely to be constrained by regulatory factors. For Lera these regulatory barriers to convergence stem from differences in the technical and service characteristics and in the development dynamics of the two sectors. In his analysis of these differences it is not always clear whether he is discussing only the Spanish case or is attempting to generalise more widely. If the latter is the case he has perhaps not taken sufficient account of national specificities. For instance in the UK broadcasting, cable and telecommunications are all regulated at the national level, if cable TV operators offer telecommunication services their tariffs are regulated in the same way as telecom operators and so on.

Hanada reviews recent developments in regulatory and policy development in Japan in relation to the problems thrown up by convergence and concluded that convergence is bringing about a total transformation of the transmission order which will require a totally new look at regulation. This new look will need to fundamentally reconsider the relationship between four factors; network operator, network user, a regulatory body and the audience. Hanada argues that this fundamental re-examination has not been begun in

Japan because the policy process is dominated by existing interests.

From these two papers I want now to draw out some key common themes for discussion.

Themes for further discussions

First the question of common carriage. Here the question everywhere is should the dominant telecommunications network operator be allowed to enter the cable TV market and if so under what terms? Should the traditional telecommunications model of separating control of transmission from control of content be universalised in a broadband environment and if so should it be applied to all forms of transmission, i.e. satellite and terrestrial radio transmission as well as cable?

Hanada calls this the choice between the separation and unification model. He associates the unification model with broadcasting, although in fact, while this holds for Japan and the UK it is by no means a universal model. There is thus a subsidiary question as to the extent to which distinct national traditions in this area influence development paths.

This question is posed differently in countries which have introduced telecommunications network competition and those which haven't, since in countries where network competition is seen as desirable cable TV networks may be seen as the basis for this competition and so have to be protected against a dominant telecommunications operator.

Second the question of content regulation. In a broadband environment can any regulation of content be justified and if so according to what principles? Hanada argues, and I agree, that there is a case for ensuring that certain universal public broadcasting services continue to be made available to everyone. On the other hand telecommunication based information services have recently produced calls for content regulation in both France and the UK. Questions of protection of privacy rights in respect of junk Faxs is a current regulatory issue in the UK. This issue has been raised in the US in terms of whether cable network operators are or are not protected from content reguation by the 1st Amendment. It is almost certain that it will be impossible to separate this issue from that of competition policy i.e. whether information services are regulated may depend upon their market share.

Third the question of interoperability. What regulatory steps, if any, need to be taken to ensure that developments in both the telecommunications and cable TV networks take place in such a way that even if we migrate towards a world of many competing networks, interoperability between these networks can be ensured?

This question clearly relates to that of common carriage. One way of posing the question is to ask what an ONA or ONP for broadband services would look like and upon whom should it be imposed.

Fourth the question of tariff policy. In brief the question here is how consumers are protected and fair competition is ensured between those offering service, while at the same time reconciling the very different tariffing methods, related to distinct service and market characteristics, adopted for broadcasting and telecommunications? Will the pricing structure for broadband be driven by the economics of TV delivery (large quantities of bandwidth and long connect times) with other traffic carried at the margins and if so what implications does this have for the economics of integration?

Fifth the question of efficient utilisation of spectrum. As the demand for mobile communication grows so too does the pressure to move broadcasting with its extremely heavy use of spectrum off the airwaves. How big a factor is this likely to be in influencing policy makers to positively favour broadband development?

Sixth the question in what ways is the development of convergence altering the structure of the broadcasting industry. Hanada argues that it is leading to even greater capital concentration and its take over by large industrial capitals. Is this generally true?

Finally I think that all three papers point to the importance of analyzing national differences in approach and development. Are the splits between Ministry responsibilities and related client groups common to other countries besides Japan and the UK and if so how will that influence development. How important to the shape of development is the relative maturity of the cable TV industry? Are countries with network competition disadvantaged in the move to broadband or on the contrary will such competition create demand to pull through development?

List of Contributors

Numbers in parentheses indicate the pages on which the authors' contributions begin.

Alleman, James H.; (129), International Center for Telecommunications Management, University of Nebraska, Omaha, Nebraska 68182, USA

Allen, David; (187), Center for Technology, Policy and Industrial Development, Massachusetts Institute of Technology (MIT), Cambridge, Massachusetts 02139, USA

Cave, Martin; (57), Faculty of Social Sciences, Brunel University, Uxbridge, Middlesex UB8 3PH, Great Britain

Chamoux, Jean Pierre; (307), Association Droit & Informatique, 75015 Paris, France

Dang Nguyen, Godefroy; (85), Ecole National Supérieure de Telecommunication (ENST) Bretagne, 29285 Brest, France

Delcros, Bertrand; (307), Chargeurs SA, 75008 Paris, France

Elixmann, Dieter; (3), Wissenschaftliches Institut für Kommunikationsdienste (WIK), 5340 Bad Honnef, West-Germany

Garnham, Nicholas; (319), Centre for Communication and Information Studies (CCIS), Polytechnic of Central London, London WC1V 7DN, Great Britain

Gassmann, Hans-Peter; (161), OECD, 75775 Paris 16, France

Hanada, Tatsuro; (277), Research Institute of Telecom-Policies and Economics (RITE), Tokyo 107, Japan

Lera, Emilio; (265), Price Waterhouse, Telecommunications Division, Madrid, Spain

Mansell, Robin; (19), Centre for Information and Communication Technologies, Science Policy Research Unit, University of Sussex, Falmer Brighton East Sussex BN1 9RF, Great Britain

Nambu, Tsuruhiko; (41), Faculty of Economics, Gakushuin University, Tokyo 171, Japan

Nyevrikel, Emilia; (221), Research Institute of the Hungarian PTT (PKI), Budapest IX Zombori U2, Hungary

Olsen, Ole Jess; (195), Institute of Economics and Planning, Roskilde University, 4000 Roskilde, Denmark

Rappoport, Paul N.; (129), Temple University, Philadelphia, Pennsylvania 19122, USA

Scherer, Joachim; (235), Kanzlei Döser, Amereller, Noack, Frankfurt, and Frankfurt University, 6000 Frankfurt, West-Germany

Schnöring, Thomas; (95), Wissenschaftliches Institut für Kommunikationsdienste, 5340 Bad Honnef, West Germany

Stanley, Kenneth B.; (129), Federal Communications Commission, Washington D.C. 20554, USA

Trotter, Stephen; (57), School of Economic and European Studies, University of Hull, Hull HU6 7RX, Great Britain

Wieland, Bernhard; (165), Faculty of Social Science, University of Cologne, 5000 Cologne, West-Germany

Wigglesworth, W.R.B.; (257), OFTEL, London EC1N 2HQ, Great Britain

GPSR Compliance

The European Union's (EU) General Product Safety Regulation (GPSR) is a set of rules that requires consumer products to be safe and our obligations to ensure this.

If you have any concerns about our products, you can contact us on ProductSafety@springernature.com

In case Publisher is established outside the EU, the EU authorized representative is:

Springer Nature Customer Service Center GmbH
Europaplatz 3
69115 Heidelberg, Germany

Batch number: 10844827

Printed by Printforce, the Netherlands